Exocrinology

Charles F. Streckfus

Exocrinology

A Textbook and Atlas of the Exocrine Cells, Glands and Organs

 Springer

Charles F. Streckfus
Department of Diagnostic and Biomedical Sciences
University of Texas School of Dentistry at Houston
Houston, TX, USA

ISBN 978-3-030-97554-8   ISBN 978-3-030-97552-4   (eBook)
https://doi.org/10.1007/978-3-030-97552-4

This Springer imprint is published by the registered company Springer Nature Switzerland AG
The registered company address is: Gewerbestrasse 11, 6330 Cham, Switzerland

To My late brother William and to my sister,
Linda, who took care of him during his
protracted illness.
My wife Cynthia
My son and daughters
My mentor Bruce J. Baum, DDS, PhD
My relatives both past and present
My colleague Lenora G. Bigler, PhD

Preface

There are numerous books on endocrinology, which academicians and students use daily; however, this is not the case for the science of exocrinology. Consequently, the purpose of this book is to fill the void and provide the readers with a comprehensive overview of the exocrine system.

The book utilizes a systems approach to exocrinology. For example, the introduction provides basic information concerning the difference between the endocrine and exocrine glands. It relates the variability of exocrine glands according to shape, method of secretion, and their secretory products.

Considering that exocrine glands are ubiquitous throughout the body ranging from the scalp to the soles of the feet, the author proceeds to present the information according to organ systems. The book starts with integument (outside covering) and then continues internally to the respiratory system, digestive tract, sensory organs, urinary and reproductive systems. The book also covers exocrine system of varying complexity ranging from exocrine cells (goblet cells) to glands (lacrimal glands) to exocrine organs (exocrine pancreas).

The book requires a background in histology; however, the author has written the chapters so that are easy to comprehend and has provided the reader with numerous illustrations making it easier to comprehend the subject matter. Taken together, the author hopes to educate the reader with material while making the process a pleasurable learning experience.

Houston, Texas, USA   Charles Streckfus

Acknowledgements

I would like to thank University of Texas School of Dentistry at Houston for their support.

The author would like to thank the following individuals for allowing the author to use their medical images and histological slides.

David G. King, Southern Illinois University, Carbondale, Illinois

Richard A. Bowen, Colorado State University, Fort Collins, CO 80523

James Edgar, Cambridge University, Cambridge CB2 1TN, United Kingdom

Peter Takizawa, Yale University, New Haven, Connecticut

Stephen Gallik, University of Mary Washington, Fredericksburg, VA 22401

Barry Rittman, University of Texas School of Dentistry at Houston, Houston, Texas

Todd Clark Brelje, University of Minnesota, Minneapolis, MN 55126

Theresa Kristopaitis, Kelli A. Hutchens, Loyola Stritch School of Medicine, Mayfield, Illinois

Pearson Scott Foresman, Publisher.

Lutz Slomianka, The University of Western Australia, Crawley, WA 6009, Australia

wikimedia.org/wikipedia/commons/d/d0/Three_Main_Layers_of_the_Eye.png

"Medical gallery of Blausen Medical 2014". WikiJournal of Medicine 1 (2). DOI:10.15347/wjm/2014.010. ISSN 2002-4436.

Contents

Abbreviations

BPH	Benign prostate hyperplasia
C6	Cervical vertebrae 6
CC16	Club cell secretory protein 16
EGF	Epidermal growth factor
GI	Gastrointestinal
HCl	Hydrochloric acid
HLA-DR	Human Leukocyte Antigen DR isotype
ME	Myoepithelial
MHC	Major Histocompatibility Complex
MUC1	Mucin 1 Cell Surface Associated
MUC2	Mucin 2 Cell Surface Associated
MUC4	Mucin 4 Cell Surface Associated
MUC5AC	Mucin 5AC, Oligomeric Mucus/Gel-Forming
MUC7	Mucin 7 Cell Surface Associated
PSA	Prostate specific antigen
SA	Secretory acini
sIgA	Secretory immunoglobulin A
SPID	Spiral intraepidermal duct
STID	Straight intraepidermal duct
SV	Seminal vesicles
T7	Thoracic vertebrae 7

List of Figures

List of Tables

Introduction to Exocrinology

1

Abstract

This chapter explains the basic knowledge of glandular tissues. It describes the difference between exocrine and endocrine glandular tissues by explaining their respective tissue origins, structure, and secretory products. It also recounts how endocrine and exocrine glands vary in cell number, size, shape, type of secretion, and branching pattern.

Learning Objectives
After reading the chapter, the reader should know the following concepts:

1. The definition of glands and glandular tissue.
2. The difference between exocrine and endocrine glands.
3. The different types of exocrine glands.
4. How endocrine glands vary in cell number, size, shape, type of secretion, and branching pattern.

1.1   Glands

Glands by definition are composed of distinct types of cells, which are specialized to produce substances to be used elsewhere in the body. These cells are known as the glandular epithelium. These cells form an aggregate and develop into an organized structure for secretion or excretion. The aforementioned aggregations are known as glands. Glands are classified according to their secretory system. In the human body, there are two types of glands: *endocrine* and *exocrine*. The study of endocrine glands is called *endocrinology* and the study of the exocrine glands is called *exocrinology*.

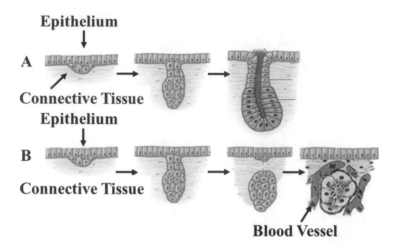

Fig. 1.1 A diagrammatic representation of the development of exocrine (**a**) and endocrine (**b**) glands. As illustrated, the endocrine gland loses its duct and becomes engorged with arteries

Endocrine glands are composed of specialized epithelial cells that secrete their products directly into the blood or lymph. They release their products basally, which allows their secretions to go through the basal lamina, move into the underlying connective tissue, and enter the vascular system. The glandular endocrine system includes the pineal gland, pituitary gland, pancreas (islets of Langerhans), ovaries, testes, thyroid gland, parathyroid gland, hypothalamus, gastrointestinal tract, and adrenal glands.

The exocrine glands originate from a layer of epithelia similar to that of the endocrine glands; however, instead of "pinching-off" and separating from the epithelial layer as in the case of the endocrine glands, they develop a duct or a system of ducts onto the apical or epithelial surface (Fig. 1.1). The signature feature of the exocrine gland is that they open to the surface (e.g., the sweat glands) or into a lumen, which eventually connects to the body surfaces such as the gastrointestinal tract or the lungs. The major glands of the exocrine system are the lacrimal glands, mammary glands, sweat glands, salivary glands, the exocrine pancreas, the kidney, and the liver; however, there is a multitude of minor exocrine glands throughout the body.

1.2   Types of Exocrine Glands

The exocrine glands, in general, have a secretory end piece, which produces the secretion, and a ductal system, which carries the secretion to its destination. There are several approaches to classify exocrine glands. They may be classified according to cell number, shape, type of secretion, or branching pattern. The subsequent paragraphs will attempt to describe the exocrine glands using the aforementioned criteria.

Fig. 1.2 The histology of the only unicellular exocrine gland in the human body. The goblet cell is ubiquitous throughout the body and produces hydrophilic mucinous secretions for membrane protection. P.S.C.E. in the slide represents pseudostratified columnar epithelium. Permission from Dr. Barry Ritman, University of Texas School of Dentistry

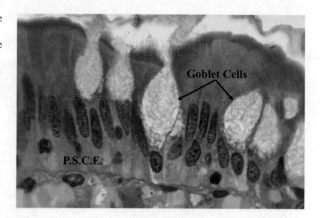

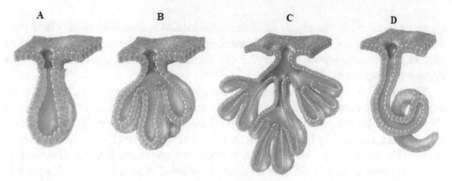

Fig. 1.3 A diagrammatic representation of the tubular type exocrine glands: a simple tubular gland (**a**), a branched tubular gland (**b**), a compound tubular gland (**c**), and a coiled tubular gland (**d**) (Adapted with permission from Quizplus.com)

With respect to cell number, there are unicellular and multicellular exocrine glands. In humans, the only unicellular gland is the *goblet cell*, which is present in the epithelial layer and releases mucins consisting of primarily glycoproteins. They are ubiquitous in the body, with the best examples located in the gastrointestinal tract and the pulmonary epithelium (Fig. 1.2).

The remaining exocrine glands in the body are multicellular. They are situated in the connective tissues beneath the epithelial layer. There they produce secretions that transit through a nonsecretory ductal system or can discharge their substances directly onto the outer epithelium.

The duct system varies in complexity and, consequently, is used to categorize the gland. The three basic types of exocrine glands are simple, branched, and compound (Fig. 1.3). Simple glands are monoductal and discharge their secretion directly to the epithelial surface.

There are two types of simple exocrine glands: the *tubular* and the *alveolar glands*. Simple straight tubular glands are very common; the best example of this

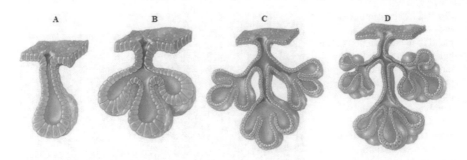

Fig. 1.4 A diagrammatic representation of the alveolar type exocrine glands in the human body: a simple alveolar gland (**a**), a branched alveolar gland (**b**), a compound alveolar gland (**c**), and a compound tubuloalveolar gland (**d**) (Adapted with permission from Quizplus.com)

type can be found in the *crypts of Lieberkühn* located within the colon. There they deliver mucinous secretions into the intestinal lumen. Simple coiled tubular glands are, as the name suggests, coiled and can be found in the scalp and function as eccrine sweat glands.

The second type is the alveolar exocrine glands. These exocrine glands can be simple, in which the secretory units directly empty into a branched excretory duct (Fig. 1.4a) or branched (Fig. 1.4b and c). These glands are generally multicellular with an elaborate, highly branched duct system that progressively empties into larger ducts. The glands have a sac-like lumen where the secretions initiate. The sac-like lumen can vary in size; if small, it is generally referred to as acinar, and if large, it is called alveoli end pieces. The salivary, lacrimal, and mammary glands and pancreas are generally referred to as compound acinar glands, while the racemose gland is an example of compound alveolar glands.

The method of secretion may also be used to classify groups of exocrine glands. There are three secretory methods among exocrine glands: *merocrine*, *apocrine*, and *holocrine*. The *merocrine* function is the most common method for producing exocrine secretions (Fig. 1.5).

Secretory cells form membrane-bound secretory vesicles internally within the cell. These vesicles migrate to the apical surface of the cell where the vesicles coalesce with the apical membrane surface and release their secretions. The salivary glands use this process to produce their secretions.

The apocrine function is a secretory process whereby the apical portion of the secretory cells is pinched-off and separated during the secretory process. The result is a secretion that has a variety of molecular components, including those of the apical membrane. An example of an exocrine gland employing this secretory process is the mammary gland (Fig. 1.6). Milk contains a myriad of large molecular components such as antibodies and lipid globules.

The holocrine function is the third secretory process. This type of secretory release involves the death of the secretory epithelial cells. The dead cells break down and their remaining constituents become their secretory product (Fig. 1.7).

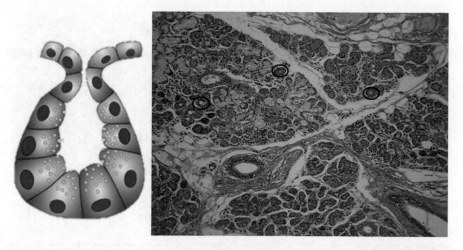

Fig. 1.5 A diagram of the merocrine function (left) and the histology of salivary gland cells (right). Individual acinar units are encircled within the diagram on the right. Courtesy of Wikipedia.com GNU Free Documentation License

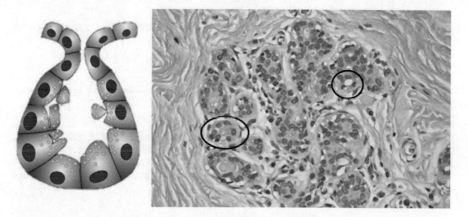

Fig. 1.6 A diagram of the apocrine function (left) and the histology of mammary gland cells (right). Individual acinar units are encircled within the diagram on the right. Courtesy of Wikipedia. com GNU Free Documentation License

The process is the most complex of the three types of secretion methods and is typically found among the sweat and sebaceous glands. Exocrine glands also vary with regard to their secretions.

The subsequent chapters of this book will describe in detail all known exocrine glands according to anatomical site, histological appearance, and function.

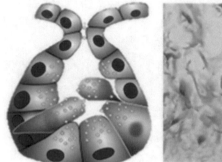 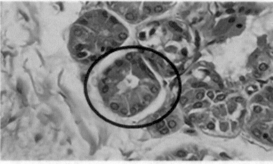

Fig. 1.7 A diagram of the holocrine function (left) and the histology of sweat gland cells (right). Individual acinar units are encircled within the diagram on the right

Questions

1. Define glands and glandular tissue.
2. Describe the difference between exocrine and endocrine glands.
3. Describe the different types of exocrine glands.
4. Describe how endocrine glands vary in cell number, size, shape, type of secretion, and branching pattern.

Suggested Readings

Gartner, Leslie. 2017. "Epithelium and Glands". In *Color Atlas and Text of Histology*, 7th ed., edited by Leslie Gartner and James Hiatt, 34–59. Philadelphia: Wolters Kluwer.
Pawlina, Wojciech. 2016. "Epithelial Tissues". In *Histology: A Text and Atlas with Correlated Cell and Molecular Biology*, 7th ed., edited by Wojciech Pawlina, 143–146. Philadelphia: Wolters Kluwer Health.

Exocrine Glands of the Integument

2

Abstract

This chapter concerns the exocrine glands of the integument. The four basic exocrine glands are the *apocrine*, *apoeccrine*, *eccrine*, and *sebaceous* glands. Apocrine, Apoeccrine, and eccrine glands are associated with perspiration, while sebaceous produces *sebum*. The chapter explains their histological characteristics, their location with respect to the body and skin layer, and their secretions.

Learning Objectives

After reading the chapter, the reader should know the following concepts:

1. The histology of the apocrine, apoeccrine, eccrine, and sebaceous glands.
2. The secretions of the apocrine, apoeccrine, eccrine, and sebaceous glands.
3. The location of these glands with respect to the human body.
4. The location of these glands with respect to the layers of the integument.

2.1 Introduction

The integumentary system is often referred to as the largest organ of the body; it covers 15–20% of the body's mass. The integumentary system consists of the skin, hair, nails, nerves, and specialized exocrine glands. Its main function is to defend the body from the environment and house the numerous internal organs within the skeletal framework. The integumentary system also functions to retain body fluids, serves as a barrier against disease, and regulates body temperature. In addition, the integumentary system works with all the other systems of the human body by rendering immunological, homeostatic, sensory, and endocrine information. Taken together, these functions maintain the internal conditions that a human body needs in order to survive.

2.2   Basic Histology of the Integumentary System

The skin portion of the integumentary system comprises the *epidermis* (ectodermal derivation) and *dermis* (mesodermal derivation) as histologically illustrated in Fig. 2.1. The epidermis is subdivided into the *stratum basale, stratum spinosum, stratum granulosum, stratum lucidum*, and the *stratum corneum*. Of particular note is the stratum basale, which gives rise to *keratinocytes* from *stem cells* located within the layer. In addition, the cells of the stratum basale give rise to a specialized apoptotic process, which produces keratin that is sloughed from the surface of the skin.

Beneath the epidermis is the dermis. The dermis is composed of dense connective tissue that provides thickness and hence increased strength and support to the skin.

Finally, there is the *hypodermis*, which is also abundantly composed of connective and adipose tissues. Adipose tissues provide insulation and energy and may be involved in androgen synthesis.

2.3   Glands of the Integument

The exocrine glands of the integument are listed in Table 2.1 and illustrated in Fig. 2.2. The four basic exocrine glands are the *apocrine, apoeccrine, eccrine*, and *sebaceous* glands. Apocrine, Apoeccrine, and eccrine glands are associated with perspiration, while sebaceous produces *sebum*.

The following paragraphs will provide a detailed description of the four types of exocrine glands.

Fig. 2.1 The histology of the integument. A represents the epidermis while B illustrates the dermis. Notice how the epidermis and the dermis interdigitate with one another. These interdigitations as related to the epidermis are called Rete pegs

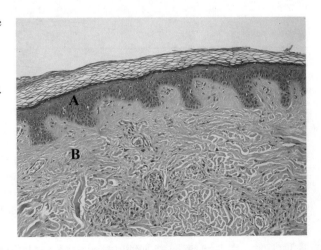

Table 2.1 Types of exocrine glands

Gland	Function	Gland type
Apocrine sweat glands	Perspiration	Coiled tubular
Apoeccrine glands	Perspiration	Coiled tubular
Eccrine sweat glands	Perspiration	Coiled tubular
Sebaceous glands	Sebum	Acinar

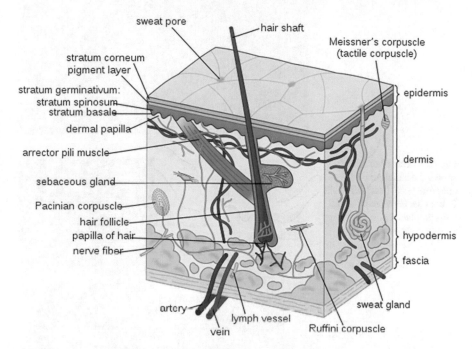

Fig. 2.2   A schematic representation of the location of the four glands to be discussed in this chapter

2.3.1   Apocrine Glands

Apocrine sweat glands are found in the armpit, areola, perineum, in the ear, and in the eyelids. The secretory portion is larger than that of eccrine glands. Rather than opening directly onto the surface of the skin, apocrine glands secrete sweat into the pilary canal of the hair follicle (Fig. 2.3).

Apocrine glands are inactive before puberty; however, hormonal changes during puberty cause the glands to increase in size and begin functioning. The glands are larger than eccrine glands and their secretions are thicker and provide nutrients for bacteria on the skin. Bacterial decomposition of apocrine sweat is what creates the acrid odor. Furthermore, the apocrine glands are adrenergically stimulated and are most active in times of stress and sexual excitement and produce pheromone-like compounds to attract other living entities within their species. Among human, studies have revealed differences between men and women in apocrine secretions and bacteria.

Figure 2.4 illustrates the structure of the apocrine gland. It is a coiled tubular gland, which originates deep within the dermis and empties into the upper portion of

Fig. 2.3 The histology of an apocrine gland. Notice that the apocrine gland (AG) empties into the hair shaft (HS) and not directly to the skin

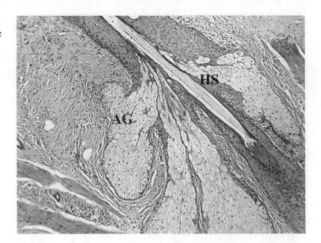

Fig. 2.4 The schematic shows the structure of the apocrine gland (left). The cells in both sections are cuboidal with the ductal portion being stratified cuboidal

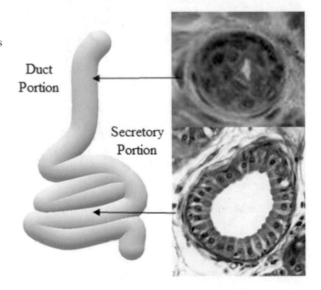

the follicular canal. The glands are divided into two sections. The secretory section is composed of simple cuboidal epithelium interposed with myoepithelial cells. The lumen is very large compared with the eccrine gland (Fig. 2.5).

The ductal portion of the apocrine gland is stratified cuboidal epithelial and somewhat resembles the ductal histology of the eccrine glands (Fig. 2.5). The lumen is narrower than the secretory portion and is usually 2–3 layers in cell thickness.

2.3.1.1 Apocrine Gland Secretions and Function

The secretions of the apocrine glands contain proteins, lipids, carbohydrates, ammonium, cholesterol, triglycerides, and numerous organic compounds. The secretion is

Fig. 2.5 The micrograph shows the "bleb-like" formations located on the apical portions of the luminal epithelia. Due to these protrusions, the gland was once thought to be apocrine in function; however, current research has demonstrated the gland is merocrine. Note the flattened myoepithelial surrounding the basal perimeter of the cuboidal epithelia

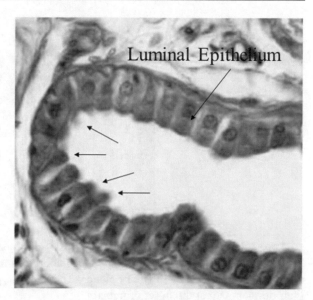

milky, proteinaceous, viscid, and colorless. Bacterial decomposition is responsible for its odor. There is a reduction in the number of glands with aging.

The function of apocrine glands is related to their location in the body. For example, apocrine sweat glands, which are associated with the presence of hair on the scalp, the armpit, and the genital region, continuously secrete a concentrated fatty sweat into the follicular tube. In the mammary gland, they contribute fat droplets to breast milk, and those in the ear help form earwax. Apocrine glands in the eyelid are sweat glands. In addition, apocrine glands in the skin are scent glands, and their secretions contain pheromones, which are involved with human behavioral and sexual interaction.

2.3.2 Apoeccrine Glands

A number of sweat glands cannot be classified as either apocrine or eccrine. As a consequence, these glands have the characteristics of both gland types and are therefore termed *apoeccrine* glands. Apoeccrine glands originate deep within the dermis and empty directly onto the surface of the skin (Fig. 2.6). With respect to size, they are smaller than apocrine glands but larger than eccrine glands. Their secretory portion is narrow and is similar to secretory coils found in the eccrine glands. However, the area of the secretory portion is wide and similar to that of the apocrine glands.

Histologically, the secretory portion is irregularly dilated. From a cellular perspective, the secretory portion of the gland has cells that resemble the clear cells of the eccrine gland and cuboidal and/or columnar cells similar to the apocrine gland.

Fig. 2.6 The micrograph presents a cross-section of an apoeccrine gland. The flattened cells on the gland's perimeter are myoepithelial (ME). The lumen is very irregular; both cuboidal (CUE) and columnar cells (CLE) surround the lumen

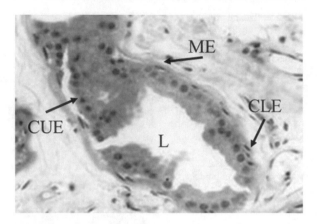

Apoeccrine glands are located in the armpits and perianal region of the body. They are presumed to have developed in puberty from eccrine glands and can comprise up to 50% of all axillary glands. They are not present at birth and begin their development at the onset of puberty.

Apoeccrine glands secrete more sweat than both eccrine and apocrine glands; thus, they play a large role in axillary sweating. Apoeccrine glands are sensitive to cholinergic activity, although they can also be activated via adrenergic stimulation. Like eccrine glands, they continuously secrete a thin, watery sweat.

2.3.3   Eccrine Glands

Eccrine sweat glands (Fig. 2.7) are ubiquitous throughout the skin except for the lips, auditory canal, prepuce, glans penis, labia minora, and clitoris. They are merocrine glands and number about 2–5 million throughout the skin. They are 10 times smaller than apocrine sweat glands and do not extend as deeply into the dermis. They excrete directly onto the surface of the skin and are involved in thermoregulation. The number of eccrine glands is age-related, with no new eccrine glands developing after birth.

The overall structure of the eccrine gland is shown in Fig. 2.7. The label SA refers to the secretory acinus or the secretory coil. If uncoiled, it would be approximately 60–80 μm in diameter and 2–5 mm in length, and the end piece of the duct would appear as an ampulla or sac-like enlargement of the duct. In this section, the ampulla or secretory fundus, there are three layers of cells: pyramidal acini or clear cells, mucoid cells, and myoepithelial cells.

Secretory acini are large, pyramidal, and clear, with the apical aspect exhibiting numerous microvilli. The cell base rests on the basal lamina or directly on the myoepithelial cells (Fig. 2.8). The nucleus is round and euchromatic.

Mucoid cells, the innermost cells in Fig. 2.8 are small and dark in appearance and are adjacent to the lumen. They mostly secrete mostly glycoproteins.

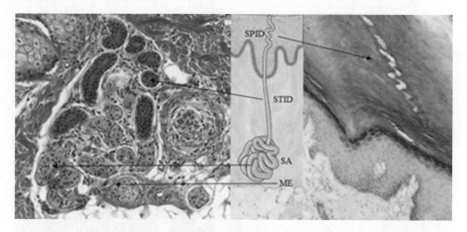

Fig. 2.7 The histology structure of an eccrine gland. The histological section on the left illustrates the structure of the straight intradermal ductal portion of the eccrine gland (STID), the secretory acinus (SA), and a myoepithelial cell (ME). The histological section to the far right shows the spiral intraepidermal duct (SPID), also known as the acrosyringium. The center diagram illustrates the location of these structures within the integument

Fig. 2.8 The histology of the cells of the ampulla or secretory fundus. Mucoid cells surround the lumen (L), followed by pyramidal clear cells (PC), and then flattened myoepithelial cells (ME)

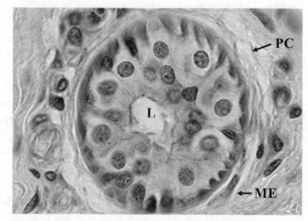

Finally, there are the myoepithelial cells, which are wedged between the basement membrane and the clear cells. They respond to cholinergic stimuli and propel the secretions to the lumen. The nuclei are flat and the cells contain numerous myofilaments. The basement membrane is peripheral to all three of these cells and separates them from the connective tissue of the dermis.

In Fig. 2.7, the straight intradermal ductal portion of the eccrine gland (STID) is located between the secretory acini and the spiral intraepidermal duct, also known as the acrosyringium (SPID). Figure 2.8 illustrates the histological structure of the STID. The three ducts shown in Fig. 2.8 comprise two layers of cells. The inner layer is the luminal cell layer, while the outer annulus of cells in the basal layer. The

Fig. 2.9 The histology of
straight intradermal ducts

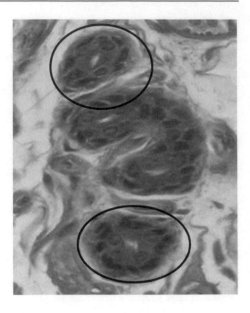

inner and outer annulus of cells are stratified cuboidal and stain darker than their
secretory counterparts stain.

A distinguishing feature from the secretory cells is the lack of myoepithelial cells
and the absence of a basement membrane. The cells are also generally small,
basophilic, and dense with mitochondria.

The spiral intraepidermal duct, also known as the acrosyringium, courses spirally
from the rete ridge to the surface of the integument. Their "cork-screw" appearance
is illustrated in Fig. 2.9. The SPID portion of the duct system can consist of an inner
luminal layer and 2–3 tiers of basal cells. The SPID cells have desmosomes and may
contain an occasional melanocyte. The mid-epidermal section to the surface is
keratinized.

2.3.3.1 The Secretions and Function of Eccrine Glands

Eccrine secretions are clear, odorless, and are composed of 98–99% water when
induced by cholinergic innervation. It was originally thought that eccrine gland
secretions primarily contain sodium chloride (NaCl), fatty acids, lactic acid, lactate,
citric acid, ascorbic acid, urea, and uric acid in a pH range from 4 to 6.8. There are
also amino acids present, mucopolysaccharides, proteolytic enzymes (kallikrein,
kininase, C1 esterase, urokinase, and cysteine proteinases), epidermal growth factor,
histamines, and ketoconazole, among many other constituents.

Current research suggests that eccrine sweat may also contain numerous biologically active constituents that are involved in the epidermal immune response. The
presence of interleukins, antibodies, and proteolytic enzymes suggests that eccrine
sweat is potentially proinflammatory. In addition, Human Leukocyte Antigen DR
isotype, which is a major histocompatibility complex (MHC) class II cell surface
receptor from the acrosyringium, may have a role in the epidermal immune response.

The literature also suggests that acrosyringeal *keratinocytes* express S100 proteins and MHC II antigens. The true importance of eccrine glands has yet to be revealed.

Eccrine glands have three primary functions. The first process is thermoregulation. Perspiration cools the surface of the skin and reduces body temperature through the process known as the latent heat of vaporization. The second function is excretion: Eccrine sweat gland secretions are an excretory route for water and electrolytes. The third function is protection. Eccrine sweat gland secretions aid in preserving the skin's acid mantle, which protects the skin from colonization by bacteria and other pathogenic organisms.

2.3.4 Sebaceous Glands

Sebaceous glands produce sebum; which is an oily, moisturizing substance with antifungal and antibacterial properties. The cells release sebum by holocrine secretion. The glands connect with the hair follicle via a short duct called the *pilosebaceous canal*. The sebaceous gland initially develops from the outgrowths of the eternal root sheath of the hair follicle.

As demonstrated in Fig. 2.10, the sebaceous duct is lined with keratinizing squamous epithelium. These cells are continuous with the lobular lipid-producing cells, i.e., the acinar cells. The acini consist of a basal layer of undifferentiated flattened epithelial cells that rest on a basement membrane. Upon differentiation, the cells become rounded and filled with lipid droplets. Their nuclei eventually shrink and are replaced by increased amounts of fat droplets. The cells ultimately burst and their contents empty into the pilosebaceous canal and onward to the infundibulum of the hair follicle.

As mentioned previously, the glandular cells release sebum by holocrine secretion. Sebum is composed of remnants of dead cells, lipids that contain triglycerides, free fatty acids, squalene, cholesterol, and wax esters.

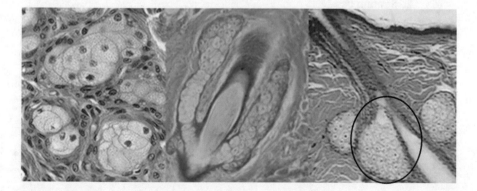

Fig. 2.10 The left micrograph shows sebaceous glands that are pear-shaped with palely stained cells (encircled on slide). The cross-sectional (center) and longitudinal (right) micrographs illustrate how the gland empties its secretion directly into the hair follicular shaft

Sebaceous glands are most closely associated with hair; however, they are found on the areola of the breast, face, eyelids, glans penis, glans clitoris, and lips. The aforementioned glands empty their contents directly to the skin. The palms of the hands and the soles of the feet lack these ubiquitous glands. Of particular interest is that sebum is the first demonstrable glandular product of the human body.

2.3.5   Other Glands

Specialized sweat glands, including the ceruminous glands, mammary glands, ciliary glands of the eyelids, and sweat glands of the nasal vestibule, are modified apocrine glands and are discussed in other chapters.

Questions

1. List the layers of the integument?
2. Name the four types of exocrine glands of the integument.
3. Which statement(s) is true about the sebaceous glands:
   a. Are associated with hairs and secrete directly onto the hair shaft.
   b. Have cells that are "foamy" in appearance.
   c. Have cells that secrete via the holocrine mode of secretion with ducts opening onto hair shaft.
   d. Secrete sebum.
   e. All of the above.
4. The integument has both eccrine and apocrine glands. Is the statement true or false?
   a. True.
   b. False.
5. Which of the following is not a layer of the epidermis:
   a. Stratum basale.
   b. Stratum spinosum.
   c. Stratum various.
   d. Stratum lucidium.
   e. Stratum corneum.
6. The apocrine glands are activated during puberty. Is the statement true or false?
   a. True.
   b. False.
7. Apoeccrine glands secrete more perspiration than both eccrine and apocrine glands; thus, they play a large role in axillary sweating. Is the statement true or false?
   a. True.
   b. False.
8. Which of the four exocrine glands is not coiled tubular in shape?

    a. Apocrine glands.
    b. Apoeccrine glands.
    c. Eccrine glands.
    d. Sebaceous glands.

Suggested Readings

Douglas Bovell. 2015. "The Human Eccrine Sweat Gland: Structure, Function and Disorders". *Journal of Local and Global Health Science*, Vol. 5. https://doi.org/10.5339/jlghs.2015.5.

Michael Ross and Wojciech Pawlina. 2006. "Integumentary System". In *Histology: A Text and Atlas*, 5th ed., edited by Michael Ross and Wojciech Pawlina, 456–474. Baltimore: Lippincott Williams & Wilkins.

Requena L., Sangüeza O. (2017) Apocrine and Eccrine Units. In: Cutaneous Adnexal Neoplasms. Springer, Cham. https://doi.org/10.1007/978-3-319-45704-8_1

Gartner, Leslie. 2017. "Chapter 11 Integument. In *Color Atlas and Text of Histology*, 7th ed., edited by Leslie Gartner and James Hiatt, 419. Philadelphia: Wolters Kluwer.

The Exocrine Glands of the Respiratory Tract

3

Abstract

This chapter discusses the exocrine glands of the respiratory tract. It will initiate the review by starting with the exocrine glands of the conducting segment and continuing through a respiratory segment of the lungs. The chapter begins with the exocrine glands of the conducting segment starting with the nasal cavity, followed by the paranasal sinuses, nasopharynx, oropharynx, larynx, trachea, bronchi, bronchioles, and concluding with the exocrine glands of the respiratory segment which are the alveolar ducts, alveolar sacs, and the alveoli.

Learning Objectives
After reading the chapter, the reader should know the following concepts:

1. The reader should have knowledge of the exocrine glands of the conducting portion which includes the nasal cavity, followed by the paranasal sinuses, nasopharynx, oropharynx, larynx, trachea, bronchi, and bronchioles.
2. The reader should have knowledge of the exocrine glands of the respiratory segment which are the alveolar ducts, alveolar sacs, and alveoli.

3.1 Introduction

The respiratory system consists of two major segments: the *conducting* and the *respiratory* portions. The conducting segment receives air into the lungs and the respiratory portion is responsible for gaseous exchange (Fig. 3.1). Upon inhalation, the air passes through the nasal cavity, where it is warmed and cleansed of gross particulates. The air then passes through the *pharynx* and into the *trachea*. The trachea bifurcates into the right and left *bronchus*. Further branching and narrowing

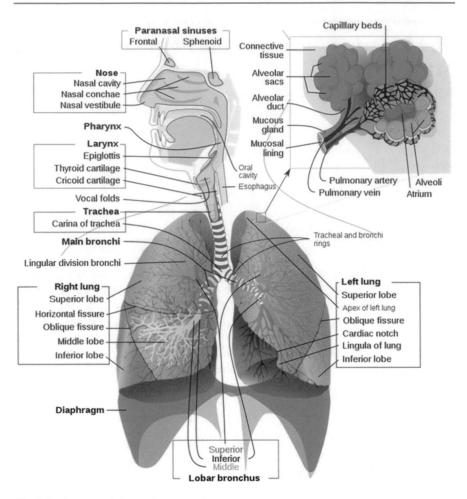

Fig. 3.1 Anatomy of the respiratory system

of the bronchi occur as the bronchus divides into numerous *bronchioles* and eventually into the *alveolar sacs* where gaseous exchanges occur.

3.2   Anatomy and Histology of the Nasal Cavity

3.2.1   Basic Anatomy

The nasal cavity is divided into two segments: the respiratory segment and the olfactory segment. The nasal cavity is the first segment of the conducting portion of the respiratory system. Anatomically, the nasal cavity is further divided into two parts by the cartilaginous *nasal septum*. Within each segment are three *nasal conchae*. The three pairs of conchae or turbinates are located on the sidewalls of

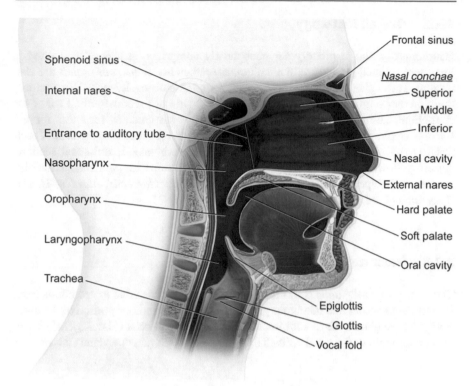

Fig. 3.2 Anatomy of the nasal cavity

the nasal cavity and serve to disrupt the airflow, directing air toward the *olfactory epithelium* on the surface of the turbinates and the septum. Located toward the posterior of the vomer or septum is the *vomeronasal organ*, the major function of which is for the detection of human *pheromones*.

Along with the nasal septum, the nasal cavity is a composite of several craniofacial bones. The base of the nasal cavities, which also form the palatal section of the mouth, is composed of the bones of the hard palate. Specifically, the horizontal plate of the palatine bone comprises the posterior base, and the palatine process of the maxilla comprises the anterior base of the nasal cavity. The anterior of the nasal cavity is the nasal vestibule and external opening, while the posterior blends, via the *choanae*, into the *nasopharynx*.

The lateral wall of each nasal cavity mainly consists of the *maxilla*; however, the perpendicular plate of the *palatine bone*, the *medial pterygoid plate*, the *labyrinth of the ethmoid*, and the inferior concha are also parts of the nasal cavity framework. The roof of each nasal cavity is formed in its upper one-third to one-half by the nasal bone. More inferiorly, the junctions of the upper lateral cartilage and nasal septum are present. Connective tissue and skin cover the components of the dorsum of the nose (Fig. 3.2).

3.2.2   Overall Histology

Histologically, the respiratory segment mostly comprises each nasal fossa and is lined with ciliated pseudostratified columnar epithelium. *Sebaceous glands* are also present. These glands aid in particle entrapment and mucosal integrity. By contrast, the olfactory segment is lined with a specialized type of pseudostratified columnar epithelium, known as the olfactory epithelium, which contains receptors for the sense of smell. This segment is located in and beneath the mucosa of the roof of each nasal cavity and the medial side of each middle turbinate. Histological sections appear yellowish-brown due to the presence of lipofuscin pigments. Olfactory mucosal cell types include *bipolar neurons*, *sustentacular cells*, *basal cells*, and *Bowman's glands*.

3.2.3   Exocrine Glands of the Nasal Cavity

The *anterior nasal glands* help moisturize the nasal mucosa due to secretions from the numerous serous and mucinous glands located in the anterior portion of the nasal cavity. These glands are in addition to ubiquitous goblet cells (Fig. 3.3), which are located throughout the nasal epithelial layer. Serous and mucous glands are present

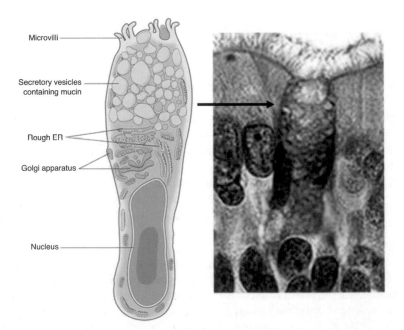

Fig. 3.3 A goblet cell with subcellular structures is illustrated in the drawing to the left. The histological slide shows the goblet cell in the middle of the slide bordered by ciliated columnar epithelia

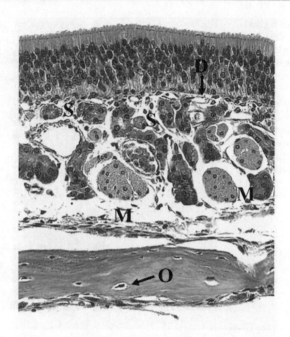

Fig. 3.4 The micrograph shows the relative size of the exocrine glands. The serous glands (S) range in size from 20.8 to 46.8 µm, while the mucinous glands (M) are 88.8 to 140.4 µm. The ductal cells (D) are 15.6 to 41.6 µm in diameter. There are no gender differences with respect to the size and number of the cells; however, the number of mucous cells decreases with age. The O in the slide shows a layer of bone with embedded osteocytes. Reproduced from Balogh K, Pantanowitz L. Mouth, nose, and paranasal sinuses. In: Mills SE, ed. *Histology for Pathologists*. third ed. Philadelphia, PA: Lippincott-Williams & Wilkins; 2007:403–430.) via Basic Medical Key, https://basicmedicalkey.com/normal-anatomy-and-histology-2/

in the *lamina propria* underlying the epithelium. The serous glands lie just beneath the lamina propria where the mucus glands are deeper within the lamina propria.

The serous acini are round in shape and smaller than the mucinous glands, which are oval and larger than their serous counterparts (Fig. 3.4). The serous glands also have a spherical, eccentrically placed nucleus, while the mucinous glands have a small, flattened nucleus, which is located near the base of the cell.

In addition, a moderate percentage of the mucosal glands are capped with *serous demilunes*, also known as *Crescents of Giannuzzi* or *Demilunes of Heidenhain*. These structures add extra quantities of *lysozyme* to assist in lowering bacterial infection in the anterior nasal cavity.

Exocrine glands are numerous throughout the nasal cavity and vary in size, number, and function. One example is the *Bowman's glands* (olfactory). They are located in the superior portion of the nasal cavity situated dorsally and caudally or proximally to the olfactory section, which is lined with specialized olfactory mucosa with a ciliated epithelium—much of which lacks goblet cells. Bowman's glands are branched tubuloalveolar and produce proteinaceous serous secretions in the lamina

Fig. 3.5 Bowman's gland. The arrow in the slide points to the Bowman's gland. The A represents the nasal cavity proper while the letter B shows the ductal portion of the gland exiting the pseudostratified, ciliated columnar epithelium. The letter C shows the connective tissue with numerous fibroblasts and blood vessels

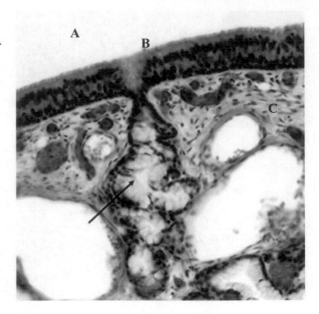

propria, where their secretions trap and dissolve odoriferous substances and detoxify gases. The major microscopic characteristic of the cells of Bowman's glands in the presence of lipofuscin granules. Lipofuscin is a pigment that produces a yellow-brown coloration to the olfactory mucosa. Electron microscopy has also revealed the presence of two types of secretory vesicles within Bowman's glands. There are large, electron-lucent vesicles found in dark cells and smaller electron-dense vesicles in light cells. Animal studies suggest that the electron-lucent vesicles contain mucous glycoproteins, while the electron-dense vesicles contain proteinaceous and serous. The ducts are relatively short and are composed of cuboidal cells (Fig. 3.5).

Bowman's gland secretions contain odorant-binding proteins that assist the olfactory region with the perception of smell. These proteins present odor chemicals to receptors of olfactory cells and elicit a smell response. In addition, the secretions can remove current chemical odorants to allow one to smell the next scent.

Bowman's gland secretions also are protective against bacterial infection because they secrete lysozyme, glycoproteins such as *aquaporin-mediated MUC5AC*, and secreted immunoglobulin A (*sIgA*). Besides the aforementioned proteins, very little is known about the exact composition or molecular identity of Bowman's gland secretions.

One major point concerning Bowman's glands is they decrease in number with age, systemic disorders, and/or infection. This phenomenon may partially explain why aged individuals have a decreased ability to smell and distinguish among types of odorants.

3.3 Paranasal Sinuses

The paranasal sinuses are also included in the respiratory system and are named according to their location within the bones of the head: maxillary, sphenoid, frontal, and ethmoid sinuses. The sinuses connect with the nasal passages, which are composed of ciliated, pseudostratified columnar epithelium, and numerous goblet cells. Both bacterial and viral infections of the sinuses are frequent maladies of the human population at large.

3.4 The Nasopharynx, Oropharynx, and Larynx

The nasopharynx and oropharynx conduct air from the nasal cavity and oral cavity to the larynx. The oropharynx is lined by stratified squamous epithelium, and the nasopharynx is lined with respiratory (pseudostratified columnar) epithelium. The nasopharynx contains seromucous glands in the lamina propria. Goblet cells are also present in the epithelial layer (Fig. 3.4).

The larynx conducts air from the pharynx to the trachea and is supported by numerous sets of complexly shaped cartilage. The larynx, in turn, is covered by three types of epithelium. Stratified squamous epithelium appears initially; it then transforms into the ciliated, stratified columnar epithelium, and finally into pseudostratified ciliated columnar epithelium (Fig. 3.6). The surface of the epiglottis is covered by nonkeratinized stratified squamous epithelium. Under the epithelium is

Fig. 3.6 The histology of the larynx. The surfaces of the larynx are lined by stratified squamous epithelium as indicated by the letter A in the figure. The lamina propria consists of loose connective tissue as indicated by the area labeled B and by exocrine glands as identified by the letter C in the figure

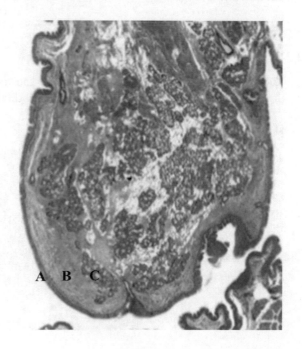

the lamina propria, which is rich in elastic fibers and numerous mixed glands. These mixed glands are both serous- and mucous-producing entities.

3.5   Respiratory System

3.5.1   Trachea

The trachea is a wide, flexible conduit, kept open by twenty C-shaped rings of hyaline cartilage. The gaps between the rings of cartilage are filled by bundles of smooth muscle (the trachealis muscle) and fibroelastic tissue. Taken together, these structures maintain flexibility and patency during inspiration and expiration.

The mucosa of the trachea is composed of the epithelium and a lamina propria. The epithelium is ciliated columnar pseudostratified epithelium resting on a well-defined basement membrane (Fig. 3.7). The aforementioned layer is infiltrated by numerous goblet cells. In addition, the supporting lamina propria under the epithelium contains elastin that plays a role in the elastic recoil of the trachea during inspiration and expiration.

The submucosa contains mixed seromucous glands. The aqueous secretions from the serous glands moisten the inspired air. Additionally, the mucus from the goblet cells, traps particles from the air. The secretions together with the epithelial cilia transport unwanted particulates upward toward the pharynx. This action keeps the lungs free from particles and bacteria. Finally, the entire organ is surrounded by *adventitia* (Fig. 3.8).

3.5.2   Intrapulmonary Bronchi

The larger bronchus is structurally similar to the tracheal mucosa except for the organization of cartilage and smooth muscle. In addition, the lamina propria contains

Fig. 3.7 The tracheal epithelium: note the numerous goblet cells (GC) and the ciliated pseudostratified columnar epithelium (PSCE). CT in the image represents connective tissue with its numerous fibroblasts

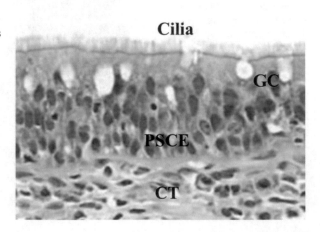

Fig. 3.8 The histology of the submucosa, which contains mixed seromucous glands. As shown in Fig. 3.8, A represents the mucosal cell layer facing the lumen of the trachea. The layer consists largely of ciliated, pseudostratified columnar epithelium. The letter B on the slide is the submucosal layer and C illustrates the mucous secreting glands

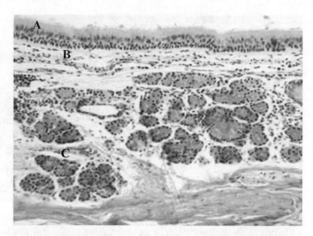

Fig. 3.9 The histology of intrapulmonary bronchi. The figure shows the mucosal layer labelled A in the diagram. It is composed of pseudostratified epithelium. The exocrine glands are labelled B. The C represents cartilage

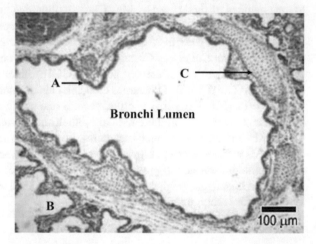

crisscrossing bundles of spirally arranged smooth muscle and elastic fibers, which become more prominent in the smaller bronchial branches. The submucosa is similar to the trachea and contains numerous small mucous and serous glands with ducts that open into the bronchial lumen (Fig. 3.9).

3.5.3   Bronchioles

Bronchioles are divided into two distinct segments: terminal and respiratory bronchioles. Both are without cartilaginous rings. The terminal bronchioles have ciliated columnar cells and some goblet cells, while the respiratory bronchioles have non-ciliated columnar cells in the smaller bronchioles. There are no goblet cells in the respiratory bronchioles, but there are club cells.

Fig. 3.10 Illustrates club cells, which are situated between the ciliated epithelium. The cell is circled in the diagram

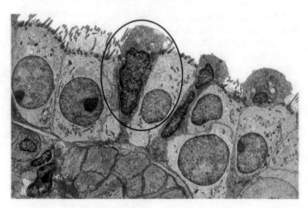

Club cells were originally called *Clara cells*, but the name has been changed due to the unethical methods used by Max Clara in obtaining human respiratory tissue. In May 2012, the editorial boards of most of the major respiratory journals including, the journals of the *American Thoracic Society*, the *European Respiratory Society*, and the *American College of Chest Physicians*, opted to change the name from Clara to club cells. Clara cell secretory protein (CC16) has been renamed to club cell secretory protein (CC16). The names club cells and bronchiolar cells are acceptable.

Club cells (Fig. 3.10) are exocrine, secrete one of the components of surfactant, and are most numerous in the cuboidal epithelium of smaller terminal bronchioles. Club cells, or exocrine bronchiolar cells, which have non-ciliated, dome-shaped apical ends with secretory granules. They secrete surfactant lipoproteins and mucins on the epithelial surface to prevent dehydration and luminal adhesion. In addition, their secretions detoxify inhaled xenobiotic compounds, antimicrobial peptides, and cytokines for local immune defense.

3.5.4   Respiratory Bronchioles

The respiratory bronchiolar mucosa is structurally similar to that of the terminal bronchioles; however, the respiratory bronchiolar mucosa has openings where gas exchange occurs. In addition, the mucosa lining consists of club cells and ciliated cuboidal cells, with simple squamous cells at the alveolar openings. Toward the ends of the respiratory bronchioles, alveoli become more numerous. The lamina is composed of smooth muscle and elastic connective tissue.

3.5.5   Alveoli

The distal ends of respiratory bronchioles branch into alveolar ducts that are lined by the openings of alveolar sacs surrounded by clusters of alveoli, both of which are lined with squamous cells. Between bordering alveoli lie thin interalveolar septa

Fig. 3.11 The histology of pneumocytes. The arrow to the right points to a type I pneumocyte where the arrow to the left features a type II pneumocyte

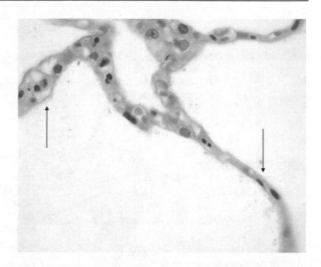

consisting of scattered fibroblasts and elastic and reticular fibers of connective tissue. There two types of alveolar cells, called *pneumocytes*, in this region of the lung (Fig. 3.11).

Type I pneumocytes are thin squamous cells that are incapable of cell division and line most of the surface of the alveoli. Type II pneumocytes are exocrine cells and produce surfactant, which is critical for airspace stability. The main constituent is a *phospholipid* called *dipalmitoylphosphatidylcholine*, which is the main surface tension reducing agent. The interalveolar septa are characterized as significantly vascularized.

3.6   Summary

As mentioned in the preceding sections, exocrine glands are ubiquitous throughout the respiratory system (Fig. 3.12). There are, however, specialized glands such Bowman's glands, goblet cells, club cells, and type II pneumocytes. Figure 3.12 presents a summary of exocrine glands and cells with respect to their anatomical site and their secretory products.

Questions

1. Which portion of the respiratory tract are the Bowman's glands located?
2. Club cells were originally called *Clara cells*, but the name has been changed due to the unethical methods used by Max Clara in obtaining human respiratory tissue. Is this statement true or false?
   a. True
   b. False
3. Goblet cells occur in all of the following except:

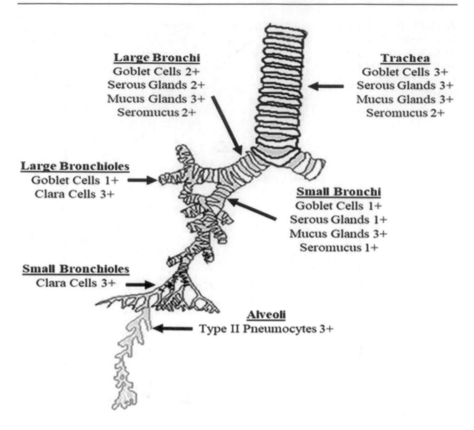

Large Bronchi
Goblet Cells 2+
Serous Glands 2+
Mucus Glands 3+
Seromucus 2+

Trachea
Goblet Cells 3+
Serous Glands 3+
Mucus Glands 3+
Seromucus 2+

Large Bronchioles
Goblet Cells 1+
Clara Cells 3+

Small Bronchi
Goblet Cells 1+
Serous Glands 1+
Mucus Glands 3+
Seromucus 1+

Small Bronchioles
Clara Cells 3+

Alveoli
Type II Pneumocytes 3+

Fig. 3.12 Location of exocrine glands within the bronchial tree. The value 3+ represents numerous cells and glands where 1+ denotes a sparsity of cells and glands. 2+ denotes a moderate presence

   a. Nasal cavity
   b. Trachea
   c. Bronchi
   d. Alveoli
4. Which of the following is MISMATCHED:
   a. Pneumocyte (alveolar cell) type I—cytoplasm through which gaseous exchange occurs
   b. Pneumocyte type II—secretes pulmonary surfactant
   c. Pneumocyte type II—stem cell for more pneumocyte types I and II
   d. Pulmonary surfactant—increases surface tension within alveoli
   e. Alveolar macrophage—removes inhaled particulates that reach the alveolus
5. Which type of cell produces surfactant? Type I or Type II pneumocytes?
6. Gaseous exchange in the alveoli occurs across the endothelial cell of the capillary, and which type of pneumocyte?

Suggested Reading

Abduxukur Ablimit , Toshiyuki Matsuzaki, Yuki Tajika, Takeo Aoki, Haruo Hagiwara, Kuniaki Takata. 2006. "Immunolocalization of Water Channel Aquaporins in the Nasal Olfactory Mucosa". *Archives of Histology and Cytology*, Vol. 69: 1–12. https://doi.org/10.1679/aohc.69.1

Boyce, John, and Shone, George. 2008. "Effects of Aging on Smell and Taste". *Postgraduate Medical Journal*, Vol. 82, No. 966: 239–241. https://doi.org/10.1136/pgm/2005039453.

Breipohl, Winrich. 1972. "Licht- und elektronenmikroskopische Befunde zur Struktur der Bowmanschen Drüsen im Riechepithel der weissen Maus". *Zeitschrift für Zellforschung und Mikroskopische Anatomie*, Vol. 131: 329–346. https://doi.org/10.1016/0014-4827(72)90497-1.

Cuschieri, Alfred and Bannister, Lawrence. 1974. "Some Histochemical Observations on the Mucosubstances of the Nasal Glands of the Mouse". *Histochemistry Journal*, Vol. 6: 543–558. https://doi.org/10.1007/BF01003270.

Frisch, Donald. 1967. "Ultrastructure of Mouse Olfactory Mucosa". *American Journal of Anatomy*, Vol. 121: 87–120. https://doi.org/10.1002/aja.1001210107.

Getchell, Marilyn, and Getchell, Thomas. 1992. "Fine Structural Aspects of Secretion and Extrinsic Innervation in the Olfactory Mucosa". *Microscopy Research and Technique*, Vol. 23: 111–127. https://doi.org/10.1002/jemt.1070230203.

Meredith, Michael. 2001. "Human Vomeronasal Organ Function: A Critical Review of Best and Worst Cases". *Chemical Senses*, Vol. 26, No. 4: 433–445. https://doi.org/10.1093/chemse/26.4.433.

Pawlina, Wojciech. 2016. "Respiratory System". In *Histology: A Text and Atlas with Correlated Cell and Molecular Biology*, 7th ed., edited by Wojciech Pawlina, 662–667. Philadelphia: Wolters Kluwer Health.

Solbu, Tom, and Holen, Torgeir. 2012. "Aquaporin Pathways and Mucin Secretion of Bowman's Glands Might Protect the Olfactory Mucosa". *Chemical Senses*, Vol. 37, No. 1: 35–46. https://doi.org/10.1093/chemse/bjr063.

Widdicombe John, and Pack, Robert. 1982. "The Clara Cell". *European Journal of Respiratory Diseases*, Vol. 63, No. 3: 202–220. https://doi.org/10.1016/S0079-6123(08)62756-9.

Winkelmann, Andreas, and Noack, Thorsten. 2010. "The Clara Cell A 'Third Reich eponym'?". European Respiratory Journal, Vol. 36, No. 4: 722–727.

Exocrine Glands of the Alimentary Tract: Section I

4

Abstract

This chapter is concerned with the exocrine glands of the upper portion of the alimentary tract namely the oral cavity. The chapter opens with the exocrine glands of the lip. It then continues to describe the glands associated with the oral mucosa, the glands of the tongue, and the three pairs of salivary glands. The chapter also describes the histology of the glands with respect to their location, type, and function. Their secretions are also described.

Learning Objectives

After reading the chapter, the reader should know the following concepts:

1. To familiarize the reader with the exocrine glands of the oral cavity. With this being said, the information contained within this chapter includes the three pairs of major salivary glands and the numerous monoductal glands of the lips, buccal mucosa, floor of the mouth, hard and soft palates, and the tongue.
2. The reader should be able to comprehend the mechanism by which the major salivary glands secrete saliva.
3. The reader should have full knowledge of the exocrine glands of the lip, tongue, and oral mucosa.

4.1　Introduction

In this chapter, the alimentary tract in this chapter is divided into three subsections. Section I concentrates on the glands of the upper digestive tract, which is located primarily within the skull. These oral glandular structures include the glands of the

© The Author(s), under exclusive license to Springer Nature Switzerland AG 2022
C. F. Streckfus, *Exocrinology*, https://doi.org/10.1007/978-3-030-97552-4_4

lip and the oral mucosa, the salivary glands, and the glands of the tongue. Section II describes the exocrine glands of the alimentary tract that extends from the foramen magnus at the base of the skull to the anus. Section III discusses the two major organs which exhibit exocrine function, the exocrine pancreas, and the liver.

4.2   Section I: The Upper Digestive Tract

4.2.1   Exocrine Glands of the Lips

The lip is divided into three histological areas. They are the *dermis*, the *vermillion border*, and the *vestibular epithelium*. Figure 4.1 illustrates the three components of the lip. The outer layer is called the *external epidermis*, which consists of an outer layer of keratinized, stratified squamous epithelia similar to the integument covering the body.

The external epidermis harbors both sebaceous and seromucous exocrine glands similar to those of the integument. In addition, the vermillion border may also contain some sebaceous glands. There are also minor salivary glands, which have mucinous secretions. The striated muscle is the *orbicularis oris*.

The *dermis* is composed of bundles of collagen and reticular fibers. Figure 4.2 illustrates the composition and the location of the exocrine glands associated with this portion of the lip.

The vermillion border is composed of keratinized, stratified squamous epithelia. This phenomenon is only found among humans. In addition, there is a paucity of sebaceous glands located in the border thus rendering the tissues susceptible to dehydration. The red color of the lips is a result of the thin, translucent epithelial layer that allows the color of the blood to appear.

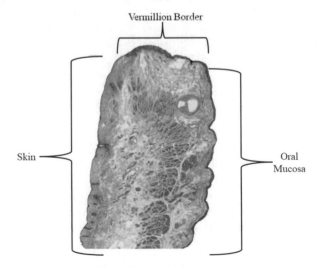

Fig. 4.1 The three components of the lip which are the dermis, the vermillion border, and the vestibular epithelium

Vermillion Border

Skin

Oral Mucosa

Fig. 4.2 The micrograph shows the three components of the external epidermis of the lip. The external epidermis contains *sebaceous* and *seromucous* exocrine glands similar to those of the integument. (**a**) Is the keratinized, stratified epithelial layer. (**b**) Is a hair follicle tubule with adjacent sebaceous glands. (**c**) Identifies the seromucus glands. Beneath the basement membrane is a collagenous matrix of numerous bundles of collagen and reticular fibers

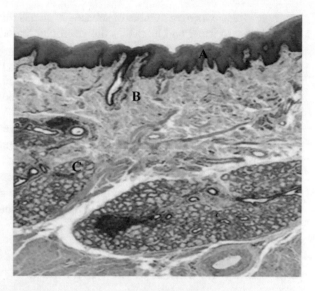

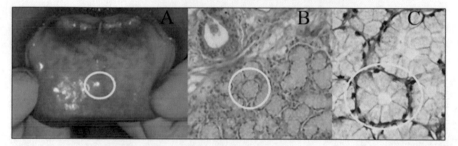

Fig. 4.3 A clinical view (designated **a**) and low-power (designated **b**) and high-power microscopic view (designated **c**) of the minor salivary glands of the vestibular portion of the oral mucosal lining. The item circled in yellow is the view of a single minor salivary gland

The third region is the vestibular epithelium. The vestibular epithelium is mucosa, which covers the lips, cheeks, alveolar mucosa, floor of the mouth, inferior surfaces of the tongue, and soft palate. The vestibular mucosa surface contains nonkeratinized stratified squamous cell epithelium. Beneath the epithelial surface is the lamina propria which consists of dense connective tissue. The submucosa contains numerous mucinous glands (Fig. 4.3). These monoductal glands produce mucin and deliver their secretions directly to the surface of the epithelial lining. As shown in Fig. 4.3, the glands present as groups of acini surrounded by connective tissue. When viewed with high power microscopy, the trapezoidal-shaped acini are bounded to one other in a circular pattern, forming a lumen in the center of the group of cells. Their nuclei are compressed toward the basement membrane due to abundant *mucinous microvesicles* located within the cell. The cytoplasm is granular and produces a "smoky" appearance when they are stained.

There is also a meager number of sebaceous glands that produce sebum, which is impermeable to water and lubricates the vestibular epithelium.

Figure 4.3 shows a clinical view (left) and low-power (center) and high-power (right) microscopic aspects of the minor salivary glands. The submucosa is comprised of abundant monoductal mucinous glands (Fig. 4.3). It is approximated that between 600 and 1000 aggregates of secretory tissue are present in the oral cavity.

As mentioned previously, the oral mucosal lining covers the lips, cheeks, alveolar mucosa, floor of the mouth, inferior surfaces of the tongue, and soft palate. The oral mucosa consists of nonkeratinized stratified squamous cell epithelium, and the lamina propria of the oral mucosa lining consists of dense connective tissue. These numerous glands are absent only on the gingiva and the anterior portion of the hard palate. The ducts are straight and they deliver the mucinous secretions directly to the surface of the epithelium.

4.2.2   Exocrine Glands of the Hard Palate

The hard palate (Fig. 4.4) is composed of keratinized stratified squamous cell epithelium; however, in some places, the epithelium can be *parakeratinized*. The lamina propria of the palatal mucosa consists of loose connective tissue, blood

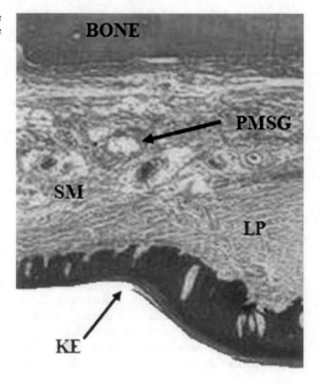

Fig. 4.4 The histology of the hard palate. KE represents the keratinized epithelium, while LP is the lamina propria. Beneath the lamina propria is the submucosa (SM), which contains numerous palatine minor salivary glands (PMSG)

vessels, and nerves. Deep in the lamina propria is a reticular layer of denser connective tissue that imbeds directly into the bone. There is no submucosa from the anterior to nearly the mesial region of the second molars. Distally, the glandular zone has a submucosa that contains adipose tissue and numerous mucous glands. The glands originate from the second bicuspid region and eventually blend into those of the soft palate. The entire mucosa is firmly attached to the palatine bones of the hard palate.

4.3   Exocrine Glands of the Tongue

The tongue is a muscular organ with a multitude of functions. The dorsal surface of the tongue encompasses a posterior area, consisting of numerous lingual tonsils situated posterior to the *sulcus terminalis*. Anterior to the sulcus terminalis is the lingual papillae that cover the remaining two-thirds of the tongue (Fig. 4.5). There are four forms of lingual papilla: filiform, fungiform, foliate, and circumvallate. With the exception of the filiform papilla, these are specialized epithelial structures responsible for the sense of taste.

Papillae are highly keratinized and covered with stratified squamous epithelium. Beneath the epithelium is a layer of loose fibrous connective tissue that forms invaginations within the layer. These tissue layers are often denoted as secondary papilla.

Fig. 4.5 The diagram illustrates the dorsum of the tongue. LT indicates the lingual tonsils, ST indicates sulcus terminalis, and MS indicates the medial sulcus. The papilla is labelled CV (circumvallate papilla), FLP (folate papilla), FGP (fungiform papilla), and FIP (filiform papilla)

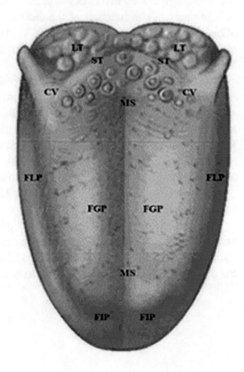

Fig. 4.6 A photo of the
clinical location of the anterior
lingual glands (ALG) and
Weber's glands (WG) on the
ventral and lateral portions of
the tongue

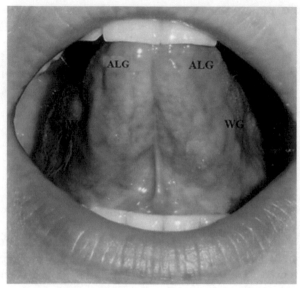

Exocrine glands of the tongue are abundant and are situated on the dorsal, lateral, and ventral surfaces of the tongue. The *glands of von Ebner* produce serous secretions and are located adjacent to the trough surrounding the foliate and circumvallate papilla. The function of the serous secretions is to cleanse the furrow and allow the taste buds to respond rapidly to varying stimuli. They also produce lipase to facilitate fat hydrolysis. Furthermore, the mucinous glands are located deep in the lamina propria of the dorsum.

The anterior lingual glands of the tongue are known as the *glands of Blandin–Nuhn*. They are located on the anterior portion of the ventral portion of the tongue and are located on both sides of the *frenulum linguae*. The glands are 12–25 mm in length and approximately 8 mm wide. The glands open to the oral cavity by three or four ducts on the ventral surface of the tongue's apex (Fig. 4.6). The saliva secretions are mixed with mucinous secretions neighboring the apex of the tongue. In contrast, the posterior portion secretions are serous/mucinous or mixed. The glands are branched tubules lined with mucous cells and topped with demilunes that provide the serous secretions.

The posterior lingual glands (*Weber's glands*) are mucinous glands located on both sides of the tongue. They can also be found in the lateral posterior portion of the dorsum of the tongue in the peritonsillar space. Weber's glands open and secrete into the crypts of the lingual tonsils. Their location is illustrated in Fig. 4.7.

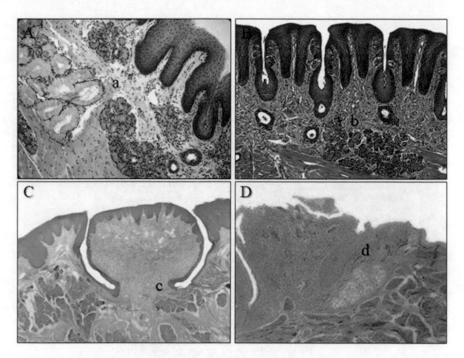

Fig. 4.7 The micrographs show the exocrine glands (Blandin and Nuhn) located in the anterior two-thirds (**a, b**, and **c**) and the posterior third (**d**) of the tongue. (**a**) The anterior tip of the tongue; "a" indicates the area of mucinous glands to the left of the letter and the serous glands below the same letter. (**b**) The location of serous glands (Weber's glands) of the foliate papillae; "b" indicates the location of these glands, which stain darker than their mucinous counterparts located in (**a**). (**c**) A circumvallate papilla with Von Ebner serous glands located beneath the letter "c." (**d**) A lingual papilla with mucinous glands below the letter "d"

4.4   The Major Salivary Glands of the Oral Cavity

4.4.1   The Parotid Gland

The *parotid gland* is a serous gland and is the largest of the major salivary glands. It weighs approximately 15–30 g and is 6 × 4 cm in dimensions. The glands are located bilaterally on the face and are positioned inferior to the *zygomatic arch* and anteroinferior to the external acoustic meatus. Additionally, they are anterior to the mastoid process and posterior to the ramus of the mandible (Fig. 4.8). The gland is divided into a base; an apex; and lateral, anterior, and posterior surfaces. Its lateral surface is situated directly beneath the skin and superficial relative to the fascia of the head. The anterior surface of the gland is grooved by the ramus of the mandible and masseter muscle. The accessory parotid gland is anterior to the parotid gland proper and posterior to *Stensen's duct*. The parotid gland has a large duct (Stensen's duct)

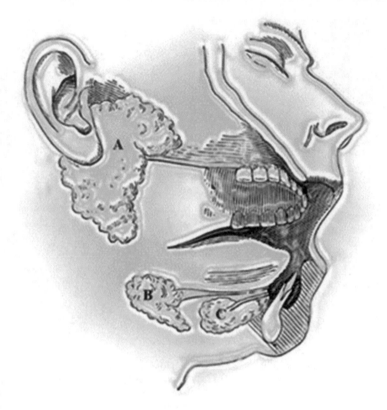

Fig. 4.8 The anatomical location of the major salivary glands: **a** is the parotid gland, **b** is the submandibular gland, and **c** is the sublingual gland

that crosses the masseter muscle and opens near the upper second molar in the oral cavity.

4.4.2   The Submandibular Gland

The submandibular gland produces both serous and mucinous secretions. It is approximately 7–10 g and 4×5 cm in size. Comparable to the parotid gland, it is also encapsulated; however, unlike the parotid gland, there is no fibrous connective tissue. The submandibular gland is located in the submandibular triangle formed by the anterior and posterior bellies of the digastric muscle and the inferior margin of the mandible (Fig. 4.8).

The gland presents with three surfaces: lateral, medial, and inferior. The gland is further divided into superficial and deep lobes, the latter representing the majority of the gland. The deep lobe projects from the mylohyoid and hypoglossal muscles. The submandibular duct also referred to as *Wharton's duct* exits from the deep part of the gland adjacent to the mandibular second molar and ascends anteriorly to the floor of the mouth (Fig. 4.8) on the summit of the sublingual papilla lateral to the lingual frenum. It is designated as the *caruncula sublingualis*.

4.4.3   The Sublingual Gland

The sublingual gland is the smallest of the major salivary glands. It is located under the mucous membrane of the floor of the mouth between the mandible and genioglossus muscle (Fig. 4.8). Unlike the aforementioned glands, the sublingual gland is not enclosed in a fascial capsule. This gland has approximately 8–20 small ducts called the *ducts of Rivinus*. These ducts exit the superior aspect of the gland and open along the sublingual fold on the floor of the mouth. Several of these ducts may occasionally join to form a common duct designated as the *duct of Bartholin*. The duct of Bartholin typically empties into Wharton's duct. The secretions from sublingual and submandibular glands flow through this duct and excrete into the floor of the mouth below the anterior portion of the tongue.

4.4.4   Salivary Gland Histology

The major salivary glands—the parotid, submandibular, and sublingual—consist of a main excretory duct, which drains into the oral cavity. Tracing the duct inward, it branches into a series of smaller ducts that progressively decrease in diameter and size. These ducts are designated as striated, intercalated, and finally into even smaller ducts termed intercellular canaliculi. These highly divided branches terminate into globular secretory end pieces known as acini (Fig. 1.6).

The salivary gland secretory end pieces contain two types of secretory cells: serous and mucous. Serous cells produce water and protein-rich content, whereas mucous cells produce mucin. This subcomponent of mucus is viscous and thus coats and protects mucosal surfaces. Salivary glands are supported by connective tissue, which houses the nerve, vascular, and lymphatic supplies. Parotid glands are mostly composed of serous cells (Figs. 4.9 and 4.10), while the submandibular and sublingual glands contain both serous and mucous cells (Fig. 4.11 and 4.12).

4.4.4.1 Histology of the Parotid Gland

Serous Acini   In the parotid glands, the secretory end pieces present themselves as an encapsulated spherical structure with 8–12 serous acini (Fig. 4.9). The acini are pyramidal in shape and rest upon a basal lamina adjacent to the connective tissue (stroma), while its narrow apex faces the central lumen (Fig. 4.9).

Serous acini in the parotid gland have a spherical nucleus, an abundance of rough endoplasmic reticulum, and a prominent Golgi complex located between the nucleus and the apex of the cell (Figs. 4.9 and 4.10). Numerous zymogen granules, where the salivary macromolecular components are stored, are generally located within the apical cytoplasm. The granules in the serous acini have a dense sphere embedded in

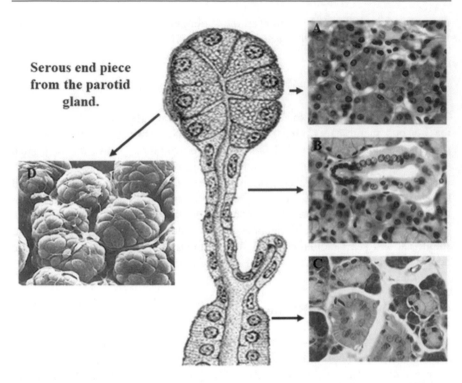

Fig. 4.9 The histology of the parotid gland secretory unit. The micrographs present pyramidal-shaped acini (**a**), intercalated ducts (**b**), striated ducts (**c**), and myoepithelial cells (**d**). Slides **a**, **b**, and **c** are cross-sectional views while slide **d** is a low-powered, lateral view of the myoepithelial cells

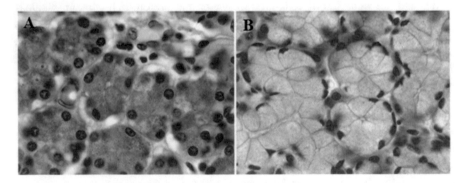

Fig. 4.10 The histological differences between serous and mucinous acini. (**a**) The serous acini (**a**) have a dark blue stippled appearance. By contrast, the mucinous acini (**b**) have a smoky appearance with flattened nuclei and a larger lumen

Fig. 4.11 A micrograph of an acinar unit, with intercellular canaliculi indicated with arrows

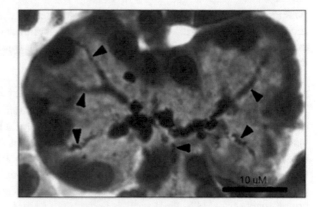

Fig. 4.12 A cross-sectional view of the excretory duct. The letter **a** illustrates the stratified pseudostratified, columnar cells of the excretory duct. **b** Shows the underlying connective tissue. Note that above the letter **b** is a blood vessel while below the **b** is a fibroblast

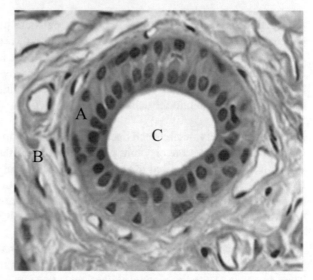

a less dense matrix. This is the hallmark feature in differentiating them from mucinous acini and the reason why they stain darker.

The basal plasma membrane at the base of an acinar cell is extensively folded to increase cellular surface area, and it interdigitates with nearby cells in folding. The apical membrane comprises intercellular canaliculi, a few rounded microvilli, and lateral membranes that have interdigitated folds. These all are membrane specialization traits involved in increasing the cell's surface area. Serous cells are linked to their neighboring cells within an acinus by an array of structures grouped as intercellular junctions, specifically termed tight junctions, adherens junctions, gap junctions, or desmosomes. A tight junction controls the passage of water and certain ions in and from the lumen to the intercellular spaces.

Adherens junctions and desmosomes mostly function to hold adjoining cells together. Gap junctions, which join the cytoplasm of adjacent cells, allow the passage of small molecules, such as ion metabolites and cyclic adenosine monophosphate (cAMP), between cells. Such cellular interchange regulates the activity of all the cells within an acinus, allowing this spherical complex to function as one unit.

Mucous Secretory Acini The mucous secretory acini are shown in Fig. 4.10b. They are predominately found in both the submandibular and the sublingual glands and *not* in the parotid gland. The cells are composed of a large number of secretory vesicles in the apical segment of the cell. These numerous vesicles crowd the cell and effectively flatten the nucleus, forcing it to the basal portion of the cell. The large Golgi complex and the endoplasmic reticula are also located at the basal area of the cell. Also, the lumen of the mucinous secretory end piece is larger than the serous counterpart.

Intercellular canaliculi Between the serous cells are prolongations of the lumen called intercellular canaliculi. They are cuboidal and often branched. In addition, they contain numerous microvilli and extend close to the basal folds (Fig. 4.11). The intercellular canaliculi and basal folds, taken together, may contribute to the formation of the fluid and electrolyte components of the primary saliva. Together they empty their contents into the acinar lumen.

Intercalated Ducts Intercalated ducts are common to all three of the major glands. Once produced, the secretory fluid from the lumen of the serous or mucinous end pieces flows into intercalated ducts (Figs. 4.9, 4.13, and 4.14). The initial cells of these ducts are directly adjacent to the secretory cells of the end pieces, and their lumina are continuous. This is unlike the intussusception of the intercalated duct with the end piece in pancreatic tissue, which will be discussed in Chap. 6.

The ducts are usually longer in the parotid compared with the submandibular glands and contribute greatly to the formation of organic content of human saliva. They are made of simple cuboidal epithelium accompanied by myoepithelial cells on their basal membrane. Intercalated ducts may serve as a reservoir of stem cells for both acini and striated ducts. These ductal cells have fewer cellular organelles compared with acinar cells. The nucleus is more oval than an acinar cell, has far less rough endoplasmic reticulum, a smaller Golgi complex, and fewer secretory granules. They are located nearer to the basal area of the cell compared with their acinar counterparts.

Several microvilli on the apical surface project into the lumen. Similar to the acinar tissue, junctional complexes such as desmosomes, gap junctions, and folded processes are involved in connecting the duct cells. The junctional complexes lack the intercellular secretory canaliculi and tend to be more apically oriented. Undifferentiated salivary gland stem cells are also thought to exist in these ducts. They can proliferate and differentiate to replace damaged or dying cells in both the secretory end pieces and the striated ducts. These cells generally secrete macromolecules such as lactoferrin and lysozyme into the salivary secretions.

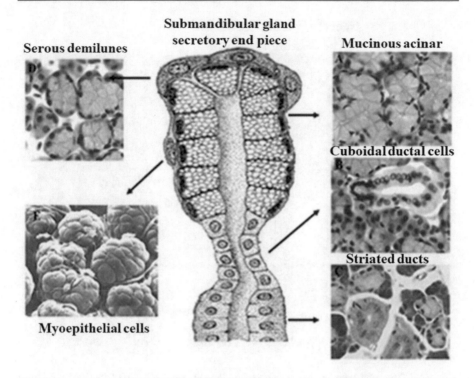

Fig. 4.13 The diagram illustrates the cellular composition of a mixed submandibular salivary gland. The cross-sectional micrographs of slides **a**, **b**, **c**, and **d** present different glandular cells types within the submandibular gland secretory end-piece. Note the presence of serous demilunes in (**d**). Slide E is a lower-powered micrograph showing the myoepithelial cells. The arrows indicate the location of the cells within the secretory end-piece

Striated Ducts Intercalated duct cells merge to form striated ducts, which constitute the largest portion of the ductal system (Figs. 4.9, 4.13, and 4.14). Striated ducts are situated within the lobules of the gland and are generally referred to as intralobular ducts. As pictured in Figs. 4.9, 4.13, and 4.14, the striated appearance at the base of these ductal cells is formed by rows of folded plasma membranes containing numerous mitochondria. This complex ductal anatomy enables these cells to participate in modifying the primary saliva by reabsorbing electrolytes, largely sodium, to formulate its hypotonic nature. They also synthesize and secrete glycoproteins such as kallikrein and epidermal growth factor. *These ducts are common only to the major salivary glands of the oral cavity.*

Excretory Ducts Striated ducts form excretory ducts also referred to as interlobular ducts, simply due to their location (between the lobules of the gland). These ducts are composed of stratified and pseudostratified columnar epithelium with numerous mitochondria (Figs. 4.9, 4.13, and 4.14). Of note, goblet cells are scattered among pseudostratified columnar epithelial cells and secrete mucin. They may be found in

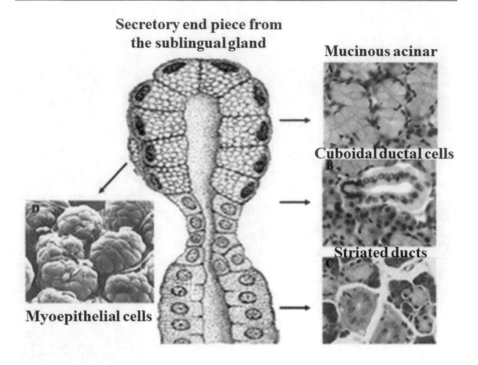

Fig. 4.14 The histology of the mucinous sublingual salivary gland. The cross-sectional micrographs of slides **a**, **b**, and **c** present different glandular cells types within the secretory end-piece. Note the presence of serous demilunes in (**c**). Slide **d** is a lower-powered micrograph showing the myoepithelial cells. The arrows indicate the location of the cells within the secretory end-piece

the excretory ducts of parotid glands, but not in the submandibular ones. These ducts also take part in the resorption of electrolytes from the saliva.

Main Excretory Ducts Smaller excretory ducts merge to form the main excretory duct. The main excretory ducts empty into the mouth, i.e., the environment. There are two important major terminal ducts: Stenson's duct of the parotid glands and Wharton's duct of the submandibular gland (Fig. 4.12). Stenson's ducts open into the mouth opposite the maxillary second molar, whereas Wharton's ducts open and deliver contents from each bilateral submandibular gland and sublingual gland to the sublingual caruncle at the base of the tongue.

Stenson's duct is lined with pseudostratified epithelium and occasional goblet cells with a layer of underlying smooth muscle cells. Wharton's duct also has pseudostratified epithelium, but unlike Stenson's duct, it includes numerous mitochondria, lysosomes, smooth endoplasmic reticula, and small vesicles. Tuft or brush cells may be present near the orifice of Stenson's and Wharton's ducts. These cells have long microvilli and apical vesicles and they are considered to be receptor cells because they have nerve endings adjacent to the basal portion of the cell. Also, dendritic cells are present and play a role in immune surveillance.

Myoepithelial Cells   Myoepithelial cells are present in both parotid and submandibular secretory end pieces (Fig. 4.13). These flat and stellate-shaped structures, with their copious branched extensions, are situated between the basal lamina and secretory cells and are linked to such cells via desmosomes (Fig. 4.13). Myoepithelial cells are epithelial in origin but have contractile functions. Their numerous extensions, filled with filaments of actin, spread out from the cell body to embrace secretory end pieces. In the presence of an appropriate stimulus, they contract and exert pressure on secretory cells, forcing them to push their content (primary saliva) from the lumen into the ductal system and eventually into the oral cavity. In addition to their role in secretion, these cells help maintain cell polarity and structural organization of the secretory end pieces by providing signaling pathways.

4.4.4.2 Histology of the Submandibular Gland
The histology of the submandibular gland is more complex than that of the parotid gland. Secretory end pieces are composed of mucous cells, which are tubular in shape. A large number of secretory vesicles, mainly containing mucus, are located in the apical cytoplasm, and thus push the nucleus, Golgi complex (relatively large), and endoplasmic reticula to the basal portion of the cell (Figs. 4.10 and 4.13). In addition, the submandibular acini are "capped" with serous demilunes, which add their secretions to the mucus-producing acini. The demilunes are further capped by the contractile myoepithelial cells. The myoepithelial cells albeit contractile are specialized epithelial cells and are related to muscle tissues. The ductal system is similar to that of the parotid glands.

4.4.4.3 Histology of the Sublingual Glands
Sublingual glands are primarily composed of mucous cells, some of which may be capped with demilunes that secrete lysozyme (Fig. 4.14). Myoepithelial cells are also present. The sublingual gland, when compared to the submandibular gland, has a greater prominence of mucous acini and fewer intralobular ducts. The ductal system is somewhat peculiar because the intercalated and striated cells are poorly developed compared with the parotid and submandibular glands. Some mucous tubules may open directly into ducts lined with cuboidal or columnar cells without typical basal striations.

Salivary Gland Function and Secretions   Whole saliva, collected from the three pairs of major glands and all the minor glands, has numerous functions: the first step in digestion (amylase), inhibition of dental caries, buffering, tooth remineralization, antimicrobial (bacteria and fungi) functions, cleansing, taste, and speech facilitation. This list is by no means exhaustive.

Secretion   Saliva is primarily secreted from the three major glands, i.e., submandibular, sublingual, and parotid. Saliva is also produced by the numerous minor salivary

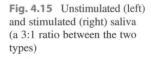
Fig. 4.15 Unstimulated (left) and stimulated (right) saliva (a 3:1 ratio between the two types)

glands found on the tongue (lingual glands), cheeks and lips (buccal and labial glands), palate (palatine glands), and on the glossopalatine folds (glossopalatine folds). Gingival crevicular fluid, a serum exudate, contributes 50 mL/day to the total whole saliva; however, it is not considered to be an exocrine secretion. Taken together, the salivary glands eventually give rise to the saliva secreted in the oral cavity. This saliva is called whole saliva because it is a composite fluid.

The salivary glands are generally activated by both the parasympathetic and the sympathetic autonomic nervous system. Parasympathetic stimulation increases the flow rate, whereas sympathetic stimulation increases salivary proteins. In addition to afferent stimuli, salivary nuclei can receive impulses from higher centers in the brain.

Saliva is produced in two flow states (Fig. 4.15): unstimulated or "resting" and stimulated. Saliva can be stimulated by mastication, the senses (smell, taste, and sight), and cognition. The constituents differ depending on the condition because some analytes are flow rate dependent. Different foods and pharmaceutical medications will produce varying flow rates. Medical conditions can also alter the salivary flow.

Whole saliva is a mixed liquid of 99% water and 1% electrolytes and protein in solution. A healthy adult produces 1.0–1.5 L of whole saliva per day. The glandular proportions are 20% from the parotid gland, 60% from submandibular, and 7–8% from sublingual glands; smaller glands in the labial, palatine, buccal, lingual, and sublingual submucosa can contribute to the remaining 10% of the overall salivary output. Upon stimulation, the parotid gland can increase its contribution to 50% of the total whole salivary secretion.

Salivary flow can be altered by circadian rhythm, circannual rhythm, hydration, smoking, medications, psychological state (e.g., fear and stress), diseases (Sjögrens syndrome), type and duration of stimulation, and aging. Likewise, salivary composition can also be altered by nutrition, hydration, smoking, diseases, and medications. The most controversial area of the aforementioned factors is the effects of aging on salivary flow. Renowned investigators have provided evidence for and against the theory of age-related changes to the salivary glands, which compromise salivary flow. However, in 1977 Scott demonstrated histologically that there are age-related changes to the salivary glands whereby the acinar cells are replaced by fibrous connective tissue. This data appears irrefutable.

Two-Step Salivary Secretion Model   Salivary production and its modification are referred to as the "two-step model" in salivary biology. The first phase is the acinar secretion of saliva and the second phase is the ductal modification of the secretion.

Acinar Cell Secretion   The accepted model for describing acinar secretions is very complex. When unstimulated, acinar cells possess a high concentration (above electrochemical equilibrium) of potassium (K^+) and chloride (Cl^-), which is a coordinated ion-exchange activity of both the Na^+/ K^+-adenosine triphosphatase (ATPase) and $Na^+/K^+/2Cl^-$ cotransporter complex.

Upon stimulation, neurotransmitters such as acetylcholine bind to receptor-specific sites on the acinar cell membrane and activate a cascade of reactions in which calcium (Ca^{2+}) plays a critical role. Once stimulated, Ca^{2+}, K^+, and apical Cl^- channels are opened, changes that allow K^+ passage from the acinar cytoplasm to the interstitial and Cl^- into the lumen. Cl^- uptake is simultaneously regulated by a second mechanism through Cl^-/HCO_3^- (bicarbonate) and Na^+/H^+ exchangers. The increase of the luminal Cl^- concentration creates an ionic gradient, which attracts Na^+ through the tight junctions to establish neutrality. As a consequence of the newly formed NaCl, an osmotic gradient moves water into the lumen through specific water channels, aquaporin 5 (AQP5), located in the apical membrane of acinar cells, and via the tight junctions of the cell. This process forms an isotonic solution, with a similar electrolyte composition to the plasma. This solution is known as primary saliva. There is a Ca^{2+} feedback system to return to a steady state in the lumen. Ca^{2+} facilitates either the degradation of inositol (1,4,5)-triphosphate (IP_3) or IP_3 receptor blockade, which may produce oscillations in the Ca^{2+} cycle.

Ductal Cell Modification of the Primary Saliva   Given that primary saliva originates from the acinar region and enters the ductal system, neural impulses stimulate specific receptors on the ductal cell surfaces to trigger a Na^+/K^+-ATPase-mediated mechanism. The signaling cascade initiates reabsorption of Na^+ and Cl^- in the ductal cell membranes followed by K^+ and HCO_3^- discharge into the lumen. Epithelial Na^+ channels, expressed in the apical membrane of salivary ducts, play an essential part in the ductal Na^+ resorption (Catalan et al. 2009). Apical Cl^- channels and Cl^-/ HCO_3^- exchangers are also hypothesized to be involved in the active resorption of Cl^- in salivary gland ducts (Peterson 1986). On the other hand, K^+ is added to the saliva in transit via K^+ channels. This ion exchange modifies the

Table 4.1 Salivary electrolytes, carbohydrates, proteins, and lipids

Electrolytes	Carbohydrates	Proteins	Lipids
Ammonia	Galactose	Apoptotic proteins	Cholesterol
Bicarbonate	Glucose	Calcium binding proteins	Diglycerides
Calcium	Mannose	Cell growth proteins	Esters
Chloride	Fucose	Cytokines	Free fatty acids
Fluoride	Glucosamine	Cytoskeleton	Glycerolipids
Hydrogen	Galactosamine	Digestive enzymes	Glycosphingolipids
Iodine	Neu5Ac	Genetic integrity proteins	Monoglycerides
Magnesium		Glycoproteins	Phosphatidylcholine
Nitrite		Histidine-rich proteins	*Phosphatidylethanolamine*
Phosphate		Immunological proteins	Phospholipids
Potassium		Membrane proteins	Sphingolipids
Sodium		Metabolic proteins	Triglycerides
Sulphates		Molecular chaperones	
Thiocyanates		Non-steroidal hormones	
		Proline-rich proteins	
		Steroidal hormones	

primary isotonic saliva to a more hypotonic fluid when unstimulated. The concentration of these salivary electrolytes is subjected to the overall salivary flow rate: When stimulated, Na^+ and Cl^- uptake decreases due to limited absorption time, thus making the saliva less hypotonic. Furthermore, the HCO_3^- concentration will increase, thus creating a more alkaline and buffered saliva during stimulation.

Salivary Protein Secretion Simply stated, secretory proteins are synthesized by ribosomes attached to the endoplasmic reticulum in acinar cells. Proteins transferred to the lumen of the endoplasmic reticulum are modified depending on function (disulfide bonds and *N*- and *O*-linked glycosylation are formed) and then translocated into the Golgi complex. Next, the proteins are further altered, condensed, and packed into secretory granules. These granules are stored in the apical cytoplasm of secretory cells. Via the process of exocytosis, there is a fusion of the granule membrane with the lumen plasma membrane of the secretory cells, followed by the rupture of the fused membranes into the lumen. Exocytosis is mainly controlled by the autonomic nervous system: Sympathetic stimulation triggers protein release from parotid and submandibular gland secretory cells.

Parotid and Submandibular Gland Secretions Each salivary gland may produce different proteins because each protein constituent may have a specific function. A broad categorical list of salivary gland components—by no means exhaustive—is provided in Table 4.1. Advances in technology have propelled biochemical research in salivary composition to the point where multiple fields of "-omics," have evolved. These advances are voluminous and beyond the scope of this chapter.

Table 4.2 Major clinically oriented salivary constituents

Component	Function
α-Amylase	Digestion
Bicarbonate	Buffering
Calcium	Remineralization
Cytokines	Cell signaling
Cytoskeleton	Cell structuring
EGF	Cell growth
Histatins	Antimicrobial
sIgA	Antimicrobial; inhibits adherence
IgM	Antimicrobial; inhibits adherence
Lactoferrin	Antimicrobial; hydrolysis of cell membrane
Lactoperoxidase	Antimicrobial; hydrolysis of cell membrane
Lipase	Digestion
Lysozyme	Antimicrobial; hydrolysis of cell membrane
Mucins	Antimicrobial; pellicle formation
Proteases	Digestion
Phosphates	Buffering
Proline-rich proteins	Antimicrobial; remineralization
Statherin	Antimicrobial; remineralization
Urea	Buffering
Water	Cleansing *solubilization*; digestion

Consequently, the following paragraphs will describe the basic clinically oriented constituents of saliva as shown in Table 4.2. Collectively, the serous acinar cells in both the parotid and the submandibular glands (demilune cells) are involved in producing proline-rich proteins (PRP), digestive enzymes, Ca^{2+}-binding proteins, and antimicrobial proteins and peptides. Most of the digestive enzymes produced by these glands, such as α-amylase, are responsible for converting polysaccharides into simpler carbohydrates in order to facilitate absorption.

The most important role of saliva is the maintenance of the oral ecology. Therefore, most of the proteinaceous constituents have antimicrobial effects in order to maintain the equilibrium between the essential oral flora and/or are involved in the remineralization of the dentition to prevent bacteria decay via acidic decalcification. The antimicrobial effects are very selective. Each microbe present in the oral cavity has a specific niche in the oral microbiome and has its own defined physiologic function. Maintaining this complex symbiotic relationship is essential for oral health. The microbial constituents work in tight synergy with one another, thereby preventing unwanted microbes from entering, finding an undesirable niche and, thereby overtaking the environment (e.g., *Candida albicans*). For example, statherin and proline-rich proteins calcium-binding proteins from the parotid and submandibular glands modulate bacterial binding to tooth surfaces by inhibiting or impeding calcium phosphate salts. This action, in turn, impedes precipitation and, thereby reduces the cariogenic activity of microorganisms of the oral flora. They are also responsible for the remineralization of hard tissues.

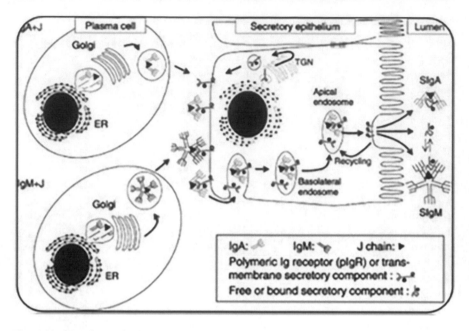

Fig. 4.16 The sIgA pathway

Besides the antimicrobial secretions of the acinar units, antimicrobial proteins can also be secreted by ductal cells. For example, lysozyme, a ductal-secreted protein, can compromise the integrity of bacterial cell walls and thus lead to bacteria breakdown in the presence of hypotonic saliva. Lactoferrin can directly or indirectly play a role in bactericidal activities in the mouth. As an iron-binding protein, it removes iron, an essential element required for bacterial survival and growth. In addition, it agglutinates bacterial cells and prevents their adherence to host cells. Moreover, it can potentially synergize the effects of lysozymes and immunoglobulins in killing bacteria.

Cystatin is another multifunctional antimicrobial protein that targets bacteria (by inhibiting bacterial proteases) and also has antiviral capabilities. Histatin, a peptide present in whole saliva, regulates oral inflammation by inhibiting the release of histamine secreted by surrounding mast cells. sIgA is the major antimicrobial protein; it is secreted by ductal cells. The antibody originates from neighboring plasma cells as a poly-Ig receptor (130 kDa) and is responsible for the uptake and transcellular transport of oligomeric (but not monomeric) IgA across the epithelial cells and into secretions such as tears, saliva, sweat, and gut fluid. Figure 4.16 diagrams show how sIgA enters saliva. Collectively, the selective oral microbiome and the associated salivary proteins maintain the homeostatic environment of the oral cavity.

Questions

1. The Von Ebner's glands are exocrine glands found on the tongue. More specifically, they are serous salivary glands that assist in the tasting of food. Are these statements true or false?
   a. True
   b. False

2. All the following statements regarding saliva buffering capacity are True EXCEPT:
   a. Bicarbonate helps to stabilize the pH and resist pH fluctuation
   b. A low pH is vital to the health of the dentition
   c. High buffering capacity resists pH changes and keeps the oral ecology stable
   d. Buffering capacity indicates how well the saliva can moderate plaque pH changes

3. Which of the following are true about salivary amylase?
   a. Foods such as bacon, beef, and fat are broken down by salivary amylase
   b. Amylase initiates digestion in the oral cavity
   c. A pH range of 2.7–4.0 is the optimum range for amylase to function as an enzyme
   d. Statements b and c
   e. Statements a and b

4. Which statement(s) are true about lingual lipase?
   a. Secreted by von Ebner's glands of tongue
   b. Hydrolyzes medium- to long-chain triglycerides
   c. Important in the digestion of milk fat in newborn
   d. Unlike other mammalian lipases, it is highly hydrophobic and readily enters fat globules
   e. All of the above

5. Which salivary analyte(s) ARE NOT involved in the mineral composition and maintenance of enamel:
   a. Calcium and Phosphate
   b. Statherin
   c. Proline-rich proteins
   d. Secretory IgA

6. Which statement(s) are true about Statherin?
   a. A phosphate-binding protein from parotid and submandibular saliva
   b. Derived from gingival crevicular fluid
   c. Amino terminal hexapeptide inhibits secondary precipitation (crystal growth)
   d. An alanine-rich protein that stabilizes inorganic ions
   e. All of the above

7. Acinar epithelium secretes proteins and hypotonic filtrate. Is this statement true or false?

a. True
b. False

8. Which statement is false concerning the submandibular gland?
   a. The submandibular salivary gland, like the parotid gland, is serous
   b. The intralobular ducts are of the same type as the parotid
   c. Ducts are more numerous and generally longer than those of the parotid gland
   d. The mucous alveoli are usually capped with serous demilunes

9. Which statement is false concerning salivary secretions?
   a. Saliva keeps the mucous membranes of the mouth moist and lubricates food for swallowing and chewing
   b. Different salivary cells secrete different proteins
   c. Parotid gland secretes a serous saliva
   d. Sublingual gland secretes mainly mucin glycoproteins
   e. None of the above

10. Parasympathetic controls increase salivary fluid secretion while sympathetic innervation of the salivary gland increases protein secretion. Is this statement true or false?
    a. True
    b. False

11. A patient is given an anticholinergic drug for the treatment of arrhythmias (irregular heartbeat). Would you expect the medication to:
    a. Increase salivary flow
    b. Decrease salivary flow
    c. Increase protein secretion
    d. No effects on either salivary flow or protein secretion

12. A patient is given an antiadrenergic drug for the treatment of hypertension. Would you expect the medication to:
    a. Increase salivary flow
    b. Decrease salivary flow
    c. Increase protein secretion
    d. Decrease protein secretion
    e. No effects on either salivary flow or protein secretion.

13. Which of the following are true about the secretion of saliva in the acinar region?
    a. The sodium pump is involved in the secretory process
    b. The cotransporter mechanism is involved in the secretory process
    c. An electrochemical gradient is created whereby sodium is transported through the tight junctions to form NaCl with the Cl ion
    d. An osmotic gradient is formed by NaCl which transports H_2O through the tight junctions
    e. All of the above

14. Passive diffusion, active transport, and endocytosis/exocytosis are all mechanisms by which particulates can enter into saliva. Is this statement true or false?

a. True
b. False

15. Which of the following salivary proteins does not have a role in host defenses?
    a. Secretory IgA
    b. Lysozymes/Proteases
    c. Lactoferrin
    d. Amylase

16. Serum and salivary IgA are exactly the same. Is this statement true or false?
    a. True
    b. False

17. Both IgA1 and IgA2 have been found in external secretions like maternal milk, tears, and saliva. Is this statement true or false?
    a. True
    b. False

Suggested Reading

Bardow, Allan, Lagerlöf, Folke, Naunalofte, Birgitte, Teneovuo, Jorma. 2009. "The Role of Saliva". In *Dental Caries: The Disease and Its Clinical Management*, 2nd ed., edited by Ole Fejerskov and Edwina Kidd, 190–207. Singapore: John Wiley and Sons.

Catalán, Marcelo, Nakamoto, Tetsuji, and Melvin, James. 2009. "The Salivary Gland Fluid Secretion Mechanism". *Journal of Medical Investigation*, Vol. 56suppl: 192–196. https://doi.org/10.2152/jmi.56.192.

Cova, Marta, Castagnola, Massimo, Messana, Irene, Cabras, Tiziana, Ferreira, Rita, Amado, Francisco, and Vitorino, Rui. 2015. "Salivary Omics". In *Advances in Salivary Diagnostics*, edited by Charles Streckfus, 63–82. Berlin: Springer-Verlag.

Fine, Daniel, Furgang, David, and Beydouin, Francis. 2002. "Lactoferrin Iron Levels Are Reduced in Saliva of Patients with Localized Aggressive Periodontitis". *Journal of Periodontology*, Vol. 73, No. 6: 624–630. https://doi.org/10.1902/jop.2002.73.6.624.

Holsinger Christopher, and Bui, Dieu. 2007. "Anatomy, Function, and Evaluation of the Salivary Glands". In *Salivary Gland Disorders*, edited by Eugene Myers and Robert Ferris, 1–16. Berlin: Springer.

Kaczor-Urbanowicz, Karolina. 2018. "Salivary Diagnostics, Salivary Glands - New Approaches in Diagnostics and Treatment, Işıl Adadan Güvenç". November 5, 2018. https://doi.org/10.5772/intechopen.73372.

Matsuo, Ryuji. 2000. "Role of Saliva in the Maintenance of Taste Sensitivity". *Critical Reviews in Oral Biology & Medicine*, Vol. 11, No. 2: 216–29. https://doi.org/10.1177/10454411000110020501.

Matczuk, Jan, Żendzian-Piotrowska, Małgorzata , Maciejczyk, Mateusz, Kurek, Krzysztof. 2017. "Salivary Lipids: A Review". *Advances in Clinical and Experimental Medicine*, Vol. 26, No. 6: 1021–1029. https://doi.org/10.17219/acem/63030.

McCoombe, Gaynor, Knox, Kenneth, and Sullivan, Hugh. 1961. "The Carbohydrate Constituents of Human Saliva". *Archives of Oral Biology*, Vol. 3, No. 3: 171–175. https://doi.org/10.1016/0003-9969(61)90134-0.

Nagato, Toshikazu, Yoshida, Hiroki, Yoshida, Aichi, Uehara, Yasuo.1983. "A Scanning Electron Microscope Study of Myoepithelial Cells in Exocrine Glands". *Cell and Tissue Research*, Vol. 209, No. 1: 1–10. https://doi.org/10.1007/BF00219918.

Nanci, Antonio. 2013. "Salivary Glands". In *Ten Cate's Oral Histology: Development, Structure, and Function*, 8th ed., edited by Antonio Nanci, 253–277. St. Louis: Mosby.

Nauntofte, Birgitte. 1992 "Regulation of Electrolyte and Fluid Secretion in Salivary Acinar Cells". *American Journal of Physiology*, Vol. 263, No. 6: 823–837.

Palk , Laurence, Sneyd James, Shuttleworth, Trevor, Yule, David, Crampin, Edmund. 2010. "A Dynamic Model of Saliva Secretion". *Theoretical Biology*, Vol. 266, No. 4: 625–640. https://doi.org/10.1016/j.jtbi.2010.06.027

Patel, Shelizeh, Barros, Juliana. 2015 Chapter 1 Salivary Glands. In: Streckfus, Charles, editor. Advances in Salivary Diagnostics. Heidelburg: Springer Press; 2015, pps. 1-16. https://doi.org/10.1007/978-3-662-45399-5.

Pawlina, Wojciech. 2016. "Digestive System I". In *Histology: A Text and Atlas with Correlated Cell and Molecular Biology*, 7th ed., edited by Wojciech Pawlina, 554–567. Philadelphia: Wolters Kluwer.

Petersen, Ole. 1986. "Calcium-Activated Potassium Channels and Fluid Secretion by Exocrine Glands". *American Journal of Physiology*, Vol. 25: G1–G13. https://doi.org/10.1152/ajpgi.1986.251.1.G1.

Pfaffe, Tina, Cooper-White, Justin, Beyerlein, Peter, Kostner, Karam, Punyadeera, Chamindie. 2011. "Diagnostic Potential of Saliva: Current State and Future Applications". *Clinical Chemistry*, Vol. 57, No. 5: 675–687. https://doi.org/10.1373/clinchem.2010.153767.

Roth, Gerald, and Calmes, Robert. 1981. "Salivary Glands and Saliva". In *Oral Biology*, edited by Robert Calmes,196–236. St Louis: Mosby.

Saracco, Charles, and Crabill, Edward. 1993 "Anatomy of the Human Salivary Glands". In *Biology of the Salivary Glands*, edited by Kathleen Dobrosielski-Vergona, 1–14. Boca Raton: CRC Press.

Scott, John. 1977. "Quantitative Changes in the Histological Structure of the Human Submandibular Glands". *Archives of Oral Biology*, Vol. 22: 221. https://doi.org/10.1016/0003-9969(77)90158-3.

Tabak, Lawrence, Levine, Michael, Mandel, Irwin, Ellison, Solon. 1982. "Role of Salivary Mucins in the Protection of the Oral Cavity". *Journal of Oral Pathology,* Vol. 11: 1–17. https://doi.org/10.1111/j.1600-0714.1982.tb00138.x.

Tiwari, Manjul. 2011. "Science Behind Human Saliva". *Journal of Natural Science, Biology, and Medicine*, Vol. 2, No. 1: 53–58. https://doi.org/10.4103/0976-9668.82322.

Turner, James, and Sugiya, Hiroshi. 2002. "Understanding Salivary Fluid and Protein Secretion". *Oral Diseases*. Vol. 8, No. 1: 3–11. https://doi.org/10.1034/j.1601-0825.2002.10815.x

Va Valenti, Piera, Berlutti, Francesca, Conte, Maria Pia, Longhi, Catia, Seganti, Lucilla. 2004. "Lactoferrin Functions: Current Status and Perspectives". *Journal of Clinical Gastroenterology*, Vol. 38, No. 6: S127–S129. https://doi.org/10.1097/01.mcg.0000128941.46881.33

Walz, Anke, Stühler, Kai, Wattenberg, Andreas, Hawranke, Eva, Meyer, Helmut, Schmalz, Gottfried, Blüggel, Martin. 2006. "Proteome Analysis of Glandular Parotid and Submandibular-Sublingual Saliva in Comparison to Whole Human Saliva by Two-Dimensional Gel Electrophoresis". *Proteomics*. Vol. 6, No. 5: 1631–1639. https://doi.org/10.1002/pmic.200500125

Exocrine Glands of the Alimentary Tract: Section II

5

Abstract

This chapter is involved with the exocrine glands of the middle and lower portion of the alimentary tract. The chapter opens with the exocrine glands of the esophagus and continues to the stomach, the small and large intestines to the anus and its opening. This chapter, similar to Chap. 4, illustrates the histology of the exocrine glands with respect to their location, type, and function.

Learning Objectives
After reading the chapter, the reader should know the following concepts:

1. The exocrine glands of the alimentary tract. In keeping with this objective, the reader should be knowledgeable of the exocrine glands of the esophagus, stomach, the small and large intestines, the anus and its opening to the surface of the skin.

5.1 Introduction

The alimentary tract has two subsections. Section I is the upper digestive tract, which is described in Chap. 4. The oral cavity and oropharynx are the major focus is located within the skull. Section II is the gastrointestinal tract (Fig. 5.1), which extends from the base of the skull to the anus. The entire length of the alimentary canal varies from location to location, but it has the same structural organization throughout its length, consisting of the mucosa, lamina propria, muscularis mucosae, submucosa, muscularis externa, and the serosa. The adventitia may be present in some instances and is usually attached to retroperitoneal organs or the body wall.

Fig. 5.1 The alimentary
canal. This work has been
released into the public
domain by its author, Pearson
Scott Foresman

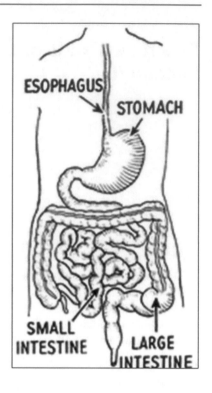

5.2   Esophagus

The esophagus is at the beginning of the alimentary canal; it is a muscular tube that is
20–35 cm in length. It starts at the C6 vertebra inferior to the larynx and posterior to
the trachea. It courses through the mediastinum and penetrates the diaphragm
(esophageal hiatus). It continues for another 3–6 cm where it joins the stomach
around the T7 vertebra at the cardiac sphincter.

The innermost wall of the esophagus comprises a mucosa of nonkeratinized
stratified squamous epithelia. As illustrated in Fig. 5.2, the lumen is somewhat star
shaped—a characteristic of a cross-sectional slice of an empty esophagus. These
"star-like" folds vanish upon dilation by a bolus of food.

At the *pharyngo-esophageal junction* in the lamina propria mucosae, there are
simple tubular and/or branched tubular glands. Inferior to these glands and beneath
the mucosa lies submucosa that contains *esophageal glands proper*, which secrete
mucous into the lumen (Fig. 5.2b). They are compound tubuloalveolar glands. There
are also some serous cells. The serous glands are more numerous in the lower third of
the esophagus.

In addition, there are *esophageal cardiac glands* located near the cardiac orifice
(esophagogastric junction) in the lamina propria mucosae. These glands secrete

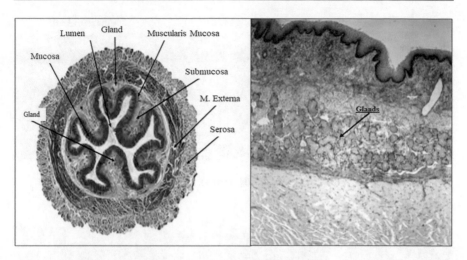

Fig. 5.2 The histology of an esophageal cross-section and its glands

mucins that protect the esophagus from acidic gastric juices. They are simple tubular or branched tubular glands.

5.3 Stomach

The *stomach* is one of several major organs of the digestive system that performs a multiplicity of exocrine functions in synchronicity with many specialized exocrine glands and cells. The organ itself resembles a J-shaped sac located inferior to the diaphragm. It has an internal volume of approximately 50 mL when empty and 1.0–1.5 L after a regular meal. Fully satiated, the volume can reach up to 4 L. Its overall function is to liquefy and partially digest the bolus of food into a soupy liquid called chyme.

The stomach has four regions: cardiac, fundic, corpus, and pyloric. Like the esophagus, the stomach has four layers. The mucosa is composed of simple columnar glandular epithelium, which in its apical regions contains mucin (Fig. 5.4). In the presence of water, these cells convert mucin into mucus. The mucosa, taken together with the submucosa, produces the *gastric rugae*. The rugae, in the presence of a full stomach, are flat and smooth; however, as the bolus moves out of the stomach and into the small intestine, they return to their original wrinkled appearance.

The inner wall of the stomach contains small pores called *gastric pits*. These pits are openings from glands that secrete hydrochloric acid (HCl) and enzymes that digest food. These glands are termed cardiac or pyloric according to their regional location. The glands that are not located in the aforementioned regions are called *gastric glands* (Fig. 5.3).

The cardiac glands are simple and/or racemose tubular mucinous structures predominately located at the esophagogastric juncture. These glands secrete a

Cardiac glands   **Gastric glands**   **Pyloric glands**

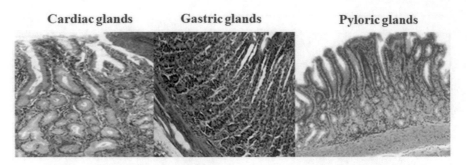

Fig. 5.3 The histology of cardiac (left), gastric (center), and pyloric exocrine (right) glands

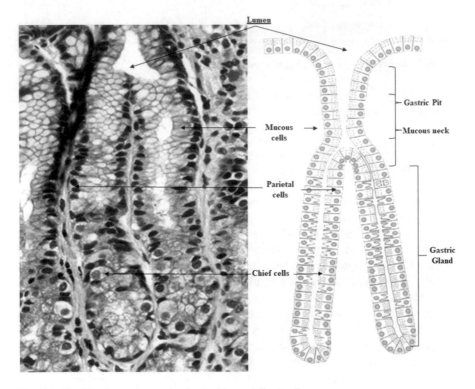

Fig. 5.4 The histology of a gastric gland with specialized cells

copious amount of mucous that protects the mucosa from the HCl secretion (Fig. 5.3). They are the fewest in the number of the three exocrine glands.

The gastric glands are branched tubular glands situated throughout the gastric mucosa except for a small area occupied by the cardiac and pyloric glands. The gastric glands have two types of specialized cells: *parietal and chief*. The parietal cells produce HCl that enters the lumen of the gastric glands, while the chief cells secrete pepsinogen. The gland also has mucinous cells, which protect the mucosa lining (Fig. 5.4).

The pyloric glands, as the name suggests, are located near the pyloric sphincter. They are branched coiled tubular glands that produce two types of mucus and the hormone gastrin.

The three types of exocrine glands vary in cell composition but, collectively, they have one or all of the cell types listed below.

5.3.1 Mucous Cells

Mucous cells are predominately located in the cardiac and pyloric glands. In the gastric glands, they are located in the neck of the gland and are called *mucous neck cells*. These cells, like those of the cardiac and pyloric glands, secrete mucous into the gastric pit.

5.3.2 Parietal Cells

Parietal cells are the exocrine cells of the stomach that secrete HCl. They are numerous and found predominately in the gastric glands, but may occasionally be present in the pyloric glands. It is extraordinary that parietal cells secrete HCl at a concentration of 160 mM at a pH of 0.8. However, due to other factors in the stomach, the pH of the stomach as a whole is 1–3. The main histological feature of the parietal cell is the presence of intracellular canaliculi.

Parietal cells are stimulated by hormones, such as gastrin and histamine, and neurotransmitters, such as acetylcholine. Parietal cells contain protein receptors on their surface for each of these activating signals. The stimulation by one type of signal by itself produces only a small amount of acid secretion, but when all three signals are combined, even at low levels, a massive secretion cascade is activated. Some medications can inhibit acid secretion by blocking the receptors of each of these three signals.

5.3.3 Chief Cells

Chief cells are the most numerous of the digestive cells, despite their absence in the cardiac and pyloric glands. Located in the gastric glands, chief cells secrete proteases to cleave proteins into amino acids and dipeptide and tripeptide chains. The main enzyme secreted by chief cells is *pepsin*. Pepsin is initially secreted as an inactive enzyme called pepsinogen. Pepsinogen becomes active when it encounters an acidic environment, whereby a peptide is removed in the presence of HCl.

Chief cells secrete digestive enzymes when they are activated by hormones and neurotransmitters. Activating hormones include secretin, vasoactive intestinal peptide, and gastrin; activating neurotransmitters include epinephrine and acetylcholine.

Secretin, vasoactive intestinal peptide, and epinephrine cause enzyme secretion in chief cells by elevating the cAMP level. Gastrin and acetylcholine cause secretion by elevating the Ca^{2+} level in chief cells. Pepsinogen secretion can be artificially blocked by drugs that antagonize the activity of these hormones and neurotransmitters.

5.4   Small Intestines

The small intestine extends from the pylorus of the stomach to the colon. It is the longest portion of the digestive tract and is divided into three parts: the *duodenum* (25 cm), the *jejunum* (2.5 m), and the *ileum* (2.5 m). The lumen of the small intestine has numerous folds known as *plicae circulares* (Fig. 5.5). In addition, the velutinous appearance of the luminal surface of the small intestine is due to the numerous finger-like projections (*villi*) of the intestinal mucosa. The villus shape differs in each part of the small intestine. For example, they are leaf shaped in the duodenum, round in the jejunum, and club shaped in the ileum. Villi extend down into the lamina propria (Fig. 5.5), where they form the crypts of Lieberkühn. Many important cells

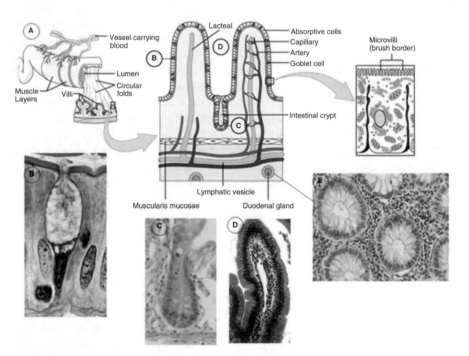

Fig. 5.5 The histology of the villi associated with the duodenum. The diagrams (**a**) present the gross and microelements of the duodenum and the cells of the intestinal crypt (crypt of Lieberkühn). The micrographs show a goblet cell (**b**), a Paneth cell (**c**), a lacteal gland (**d**), and Brunner's gland (**e**)

Fig. 5.6 Paneth cell. An individual Paneth cell is encircled by a dashed line in the figure

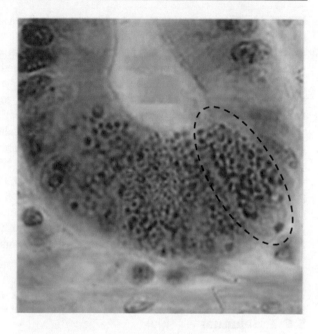

reside in the crypts (Fig. 5.5), including those involved in host defense and signaling. Stem cells that replace depleted epithelial cells in villi are also found in the crypts.

The surface of the duodenum has numerous villi, which increase the absorptive surface of the cell. Villi are composed of enterocytes that are a form of columnar epithelium. There are numerous goblet cells between the enterocytes (Fig. 5.5b).

Brunner's glands are compound tubular mucous glands in the submucosa of the duodenum and are the identifying feature of this section of the small intestine (Fig. 5.5e). Brunner's gland is a convoluted tube lined with columnar epithelium. They are embedded in the mucous coat of the duodenum; whose ducts pass inward to open on the surface of the mucous coat.

Situated between villi are simple tubular glands called the **crypts of Lieberkühn (also known as intestinal glands)**. The **crypts of Lieberkühn** contain multiple types of cells: enterocytes (absorb water and electrolytes), goblet cells (secrete mucus), enteroendocrine cells (secrete hormones), cup cells (express vimentin), tuft cells (chemosensory *cells*), and—at the base of the gland—Paneth cells (secrete antimicrobial peptides) and stem cells. Of these cell types, only the goblet and Paneth cells are exocrine in nature.

Paneth cells (Fig. 5.6) are one of the principal cell types of the small intestine epithelium. They are identified microscopically by their location just below intestinal stem cells in the crypts of Lieberkühn. These cells contain large eosinophilic granules that occupy most of their cytoplasm. These granules consist of several antimicrobial compounds and other compounds that are known to be important in immunity and host defense. Paneth cells secrete some of these compounds into their lumen in order to maintain a healthy microbial environment. This, in turn, contributes to maintaining the gastrointestinal barrier.

The principal defense molecules secreted by Paneth cells are *alpha-defensins*. These peptides have hydrophobic and positively charged domains to interact with phospholipids in cell membranes, an action that allows defensins to insert into membranes, whereby they interact with one another to form pores that disrupt membrane function, leading to cell lysis.

Paneth cells are stimulated to secrete defensins when exposed to both Gram-positive and Gram-negative bacteria and to bacterial products such as lipopolysaccharide and lipid A muramyl dipeptide.

In addition, Paneth cells secrete lysozyme, tumor necrosis factor-alpha, and phospholipase A2, all of which have clear antimicrobial activity. The aforementioned molecules provide Paneth cells with a molecular arsenal against a spectrum of pathogens which include bacteria, fungi, and some viruses. In summary, secretions from the crypts of Lieberkühn and Brunner's glands produce the **intestinal fluid** or **succus entericus.**

5.5   Jejunum

The jejunum is histologically similar to the duodenum but somewhat different with respect to the presence of exocrine glands and cells. One major difference is the absence of glands in the submucosa (Fig. 5.7). This is a signature feature of the jejunum. In addition, the jejunum has more goblet cells compared with the duodenum.

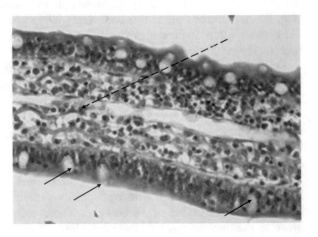

Fig. 5.7 The histology of the jejunum, with the solid arrows indicating Goblet cells intermingled with columnar epithelium. The dashed arrow points to the white blood cells located within the lacteal

5.6   Ileum

The ileum is histologically similar in structure to the duodenum and the jejunum in that it also has villi with numerous goblet cells among the columnar cells (enterocytes) and the crypts of Lieberkühn within the deeper portions of the lamina propria. In addition, the crypts of Lieberkühn have, like the jejunum, a distinct submucosa that is glandless. *Peyer's patches* are the histological hallmark of this portion of the small intestine.

5.7   Large Intestine

The large intestine is composed of the appendix, the cecum, the colon (ascending, transverse, and descending), the rectum, and the anal canal. The appendix and the anal canal will be discussed separately because the other parts of the large intestine present identical histological features.

Unlike the small intestine, the cecal and large intestinal mucosa lack villi (Fig. 5.8). Instead, the mucosa is composed of densely arranged straight, tubular glands (colonic glands). Colonic glands are lined by enterocytes and goblet cells with a much higher density of goblet cells than in the small intestine. The base (crypts) of the glands is analogous to the small intestinal crypts of Lieberkühn, and epithelial proliferation occurs in this population of epithelial cells. The rectal mucosa is similar to the large intestine in histologic appearance.

Fig. 5.8 The histology of the large intestine. Note the lack of villi and numerous goblet cells as indicated by the arrow in the figures

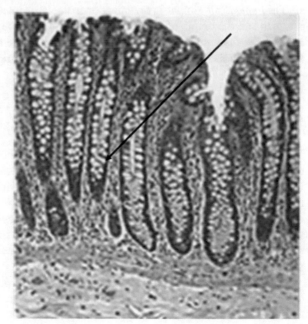

Fig. 5.9 The histology of the appendix: N, a lymphatic nodule; C, crypts of Lieberkühn

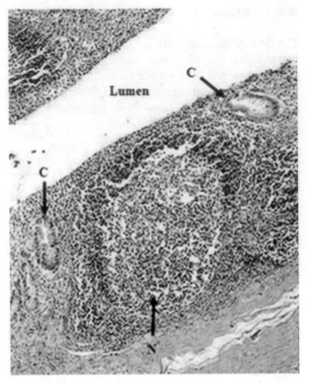

The *appendix* is similar in organization to the large intestine with regard to having a lumen and the crypts of Lieberkühn (Fig. 5.9). Of note, there are fewer crypts of Lieberkühn than the large intestine. The main difference between the two tissues is that the appendix has a lamina propria that is rich in lymphatic nodules.

5.8   Rectum and Anal Canal

When anatomically comparing the distal portion of the intestines, the rectum is distinguished by having *transverse rectal folds*. The mucosa of the rectum is similar to that of the distal portion of the large intestine because it has numerous goblet cells and straight tubular intestinal glands.

The *anal canal* is divided into three zones: colorectal, anal transitional, and squamous. The zones indicate a mucosal transition from simple columnar epithelium to stratified cuboidal epithelium to the stratified squamous epithelium of the perianal skin (Fig. 5.10).

In the colorectal zone, the exocrine microanatomy is similar to that of the rectum; there are numerous crypts of Lieberkühn that are associated with goblet cells. In addition, there are straight tubular intestinal glands or anal glands, both of which are

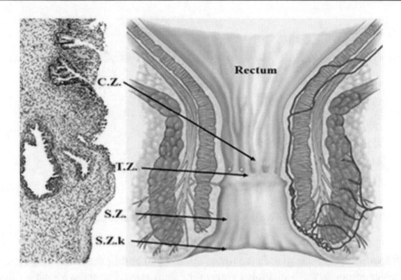

Fig. 5.10 The histology of the anal canal: C.Z., colorectal zone; T.Z., transitional zone; S.Z., squamous zone; S.Z.k., squamous zone keratinized

Fig. 5.11 The micrograph shows the presence of the anal glands located in the submucosa. They are mucous-producing glands lined with columnar epithelial cells. The cells are located below the letter A in the slide

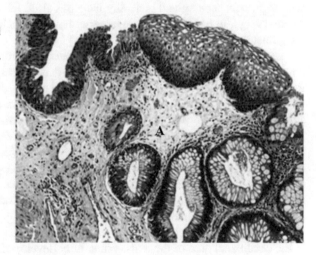

eccrine. Illustrations of these mucous glands can be seen in Figs. 5.9 and 5.10. It is worth noting that inferior to the transitional zone there are no crypts.

The transitional zone is very sparse with respect to exocrine glands. An independent sebaceous gland (without hair) may be occasionally present.

In the squamous zone (Figs. 5.10 and 5.11), there are four different types of glands: (1) holocrine sebaceous glands (2) merocrine, eccrine sweat glands (3) merocrine, apocrine scent glands, and (4) specific squamous zone eccrine glands (Fig. 5.12). These glands are generally located in the connective tissue between the sphincter ani and the epithelial lining of the anal canal.

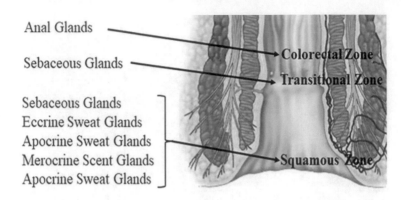

Anal Glands

Sebaceous Glands

Sebaceous Glands
Eccrine Sweat Glands
Apocrine Sweat Glands
Merocrine Scent Glands
Apocrine Sweat Glands

Colorectal Zone
Transitional Zone
Squamous Zone

Fig. 5.12 The diagram illustrates the location of the exocrine glands in the anal canal

Holocrine sebaceous glands are found circumanal and in the squamous zone and to a lesser degree in the transitional zone of the anal canal. From the transitional zone distal to the anal verge, there are three types of sebaceous glands: (1) "independent" or without hairs, (2) with rudimentary hairs, and (3) with well-developed hairs. The size of the sebaceous gland generally increases from the transitional zone distal to the anal verge. In addition, they are larger and more branched than those found in the integument.

Apocrine, merocrine scent glands are located circumanal and in the squamous zone of the anal canal. Circumanal apocrine glands resemble those of the axillary region, whereas those of the squamous zone are modified; some are simple tubules whose secretory portions are not coiled, while in others the secretory portions are ampulla-like sacs. Sweat glands are only found circumanal. Apocrine glands are nonfunctional before puberty, at which time they grow and commence secretion. The secretions contain ectohormones (pheromones), which can affect the behavior of other individuals. Humans do not have a well-developed vomeronasal organ, or Jacobson's organ, and thus the true function of these secretions is not known. The most prominent steroid hormones are 16-androstenes, androstadienone, androstenone, and androstenol.

Eccrine apocrine sweat glands produce perspiration and are predominately found in the circumanal area of the anus. When fully developed, they can appear as large balls. The main components of sweat are simple organic acids, including E-3-methyl-2-hexenoic acid and 3-methyl-3-hydroxylhexanoic acid.

5.9   Summary

A summary of the type, site, and secretions of the gastrointestinal tract is listed in Table 5.1.

Table 5.1 Summary of the exocrine glands of the digestive tract

Name of gland	Location	Secretion	Gland structure
Cobelli's glands	Esophagus	Mucus	Branched tubular
Esophageal glands	Esophagus	Mucus	Tubuloalveolar
Parietal cell	Stomach	HCl	Canaliculi
Pyloric glands	Stomach	Mucus	Compound-tubular
Chief cell	Stomach	Pepsin	Unicellular
Anal glands	Intestines	Mucus	Tubular
Anal scent glands	Intestines	Pheromones	Tubular
Anal sweat glands	Intestines	Sweat	Tubular
Brunner's glands	Intestines	Mucus	Compound-tubular
Circumanal glands	Intestines	Mucus	Sebaceous glands
Goblet cells	Intestines	Mucus	Unicellular
Lieberkühn's glands	Intestines	Mucus	Tubular
Paneth cells	Intestines	Defensins	Unicellular

Questions

1. The esophageal glands are tubular alveolar glands that secrete mucous at the superior portion of the esophagus. Is this statement true or false?
   a. True
   b. False
2. Sebaceous glands are primarily located in which part of the gastrointestinal tract?
   a. Esophagus
   b. Stomach
   c. Small intestines
   d. Large intestines
   e. Anus
3. The esophageal glands are tubular alveolar glands that secrete mucous at the superior portion of the esophagus. Is this statement true or false?
   a. True
   b. False
4. Brunner's glands are primarily located in which part of the gastrointestinal tract?
   a. Esophagus
   b. Stomach
   c. Small intestines
   d. Large intestines
   e. Anus
5. The Chief cells secrete pepsin, Is this statement true or false?
   a. True
   b. False

6. Which exocrine glands listed below are compound tubular glands?
   a. Circumanal glands
   b. Brunner's glands
   c. Pyloric glands
   d. Anal scent glands
   e. Anal sweat glands
   f. Brunner's and pyloric
7. The parietal cells secrete HCl and are located in the large intestines. Is this statement true or false?
   a. True
   b. False
8. Some anal glands produce pheromones. Is this statement true or false?
   a. True
   b. False
9. Cobelli's glands are primarily located in which part of the gastrointestinal tract?
   a. Esophagus
   b. Stomach
   c. Small intestines
   d. Large intestines
   e. Anus
10. The Lieberkühn glands secrete mucous, are tubular in structure, are found in the esophagus and are located Is this statement true or false?
    a. True
    b. False

Suggested Reading

Eglitis, John, and Eglitis, Irma. 1961. "The Glands of the Anal Canal in Man". *Ohio Journal of Science*, Vol. 61, No. 2: 65–79.

Gartner, Leslie. 2017. "Digestive System II". In *Color Atlas and Text of Histology*, 7th ed., edited by Leslie Gartner and James Hiatt, 384–415. Philadelphia: Wolters Kluwer.

Gerbe, François, Legraverend, Catherine, Jay, Philippe. 2012. "The Intestinal Epithelium Tuft Cells: Specification and Function". *Cellular and Molecular Life Sciences*. Vol. 69: 2907–2917.

Junqueira, Luis, and Carneiro, Jose. 1983. "Glands Associated with the Digestive Tract". In *Basic Histology*, 4th ed., edited by Luis Junqueira and Jose Carneiro, 342–347. Los Altos: Lange Medical Publications.

Kelly, Douglas, Wood, Richard, and Enders, Allen. 1984. "The Digestive System". In *Bailey's Textbook of Microscopic Anatomy*, 18th ed., edited by Douglas Kelly, Richard Woods, and Allen Enders, 506–565. Baltimore: Williams & Wilkins.

Nicholson, Bubba. 2011. Exocrinology The Science of Love: Human Pheromones in Criminology, Psychiatry, and Medicine. Tampa: Nicholson Science.

Ross, Michael, and Wojciech Pawlina. 2017. "Digestive System II: Esophagus and Gastrointestinal Tract". In *Histology: A Text and Atlas: With Correlated Cell and Molecular Biology*, 7th ed., edited by Wojciech Pawlina, 568–594. Philadelphia: Wolters Kluwer.

Verhaeghe, Johan, and Enzlin, Paul. 2013. "Pheromones and Their Effect on Women's Mood and Sexuality". *Facts, Views & Vision in ObGyn*, Vol. 5, No. 3: 189–195.

Exocrine Glands of the Alimentary Tract: Section III the Liver, Pancreas, and Gallbladder

6

Abstract

This chapter illustrates that there are organs of the body that have both exocrine and endocrine functions. The three major organs discussed in this section are the pancreas, the liver, and the gallbladder. The chapter will illustrate how the endocrine and exocrine units of each organ work together to produce their major secretions.

Learning Objective

After reading the chapter, the reader should know the following concepts:

1. The objective of this chapter is to present the reader with the exocrine structure of the three major organs of the alimentary tract.

6.1 Introduction

This chapter describes three major organs of the digestive tract: the pancreas, the liver, and the gallbladder, and is the final chapter on the three-part digestive system. The alimentary canal connects with the outside environment and, considering that the liver and pancreas empty their contents into the canal and eventually exit the body, they are considered to be, in part, exocrine glands. The liver and pancreas have exocrine and endocrine functions. The endocrine secretions from both glands go directly to the bloodstream, while their exocrine secretions enter the duodenum (Fig. 6.1).

The gallbladder has been included in this chapter because it contains mucinous exocrine glands at its neck. Similar to the pancreas and the liver, it delivers its contents into the duodenum and, eventually, to the outside milieu. A detailed description of the cell is presented later in this chapter.

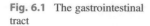

Fig. 6.1 The gastrointestinal tract

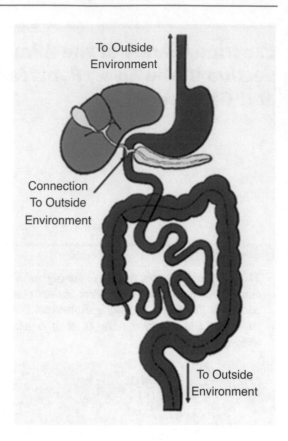

6.2   Pancreas

6.2.1   Endocrine Pancreas

The pancreas contains tissue with both endocrine and exocrine functions. The endocrine aspect of the pancreas is extremely well documented and will be briefly described. Islets of Langerhans are composed of up to 3000 secretory cells. Like most endocrine tissues, they are vascular and contain arterioles and venules that allow the hormones to be secreted directly into the systemic circulation. Microscopically, the islets of Langerhans consist of glucagon-producing α-cells (30%) and insulin-producing β-cells (60%), with the remainder made up of δ-cells (somatostatin-producing), γ-pancreatic polypeptide-producing cells, and ε-cells which produce ghrelin.

Alpha cells are on the periphery of the islet complex and secrete the hormone glucagon. Beta cells are centrally located within the islet and secrete insulin.

They are more numerous than alpha cells and found in the center of the islet. In addition, there are delta cells that produce somatostatin. The remaining endocrine cells are randomly distributed throughout the islet.

6.2.2 Exocrine Pancreas

The exocrine pancreas is classified as a compound tubuloacinar serous gland and has a significant role in food digestion. It should be noted that the dark-staining cells form clusters, which are arranged in lobes surrounded by thin walls (Fig. 6.2).

The acinar cells are very similar in structure to those of the parotid glands (Chap. 4). Figure 6.3 presents a histological image of pancreatic tissue. A cross-section through an acinus is circled and illustrates pyramidal-shaped cells that surround a small lumen.

These cells are stimulated by two hormones, namely secretin, and cholecystoki-nin, which induce the pancreas to produce chyme. Chyme contains water, ions, and

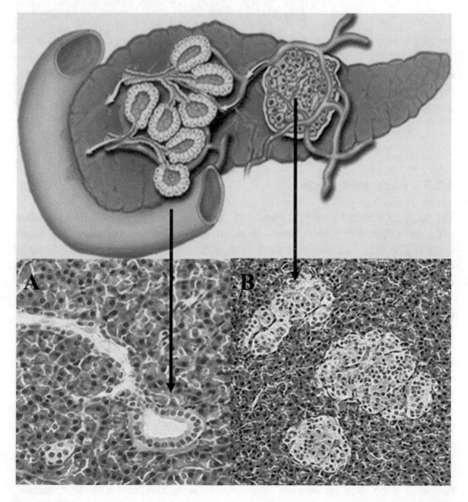

Fig. 6.2 The pancreas has exocrine and endocrine glands. The left micrograph (**a**) is an exocrine acinar unit and in the right micrograph (**b**) is an endocrine islet of Langerhans

Fig. 6.3 A micrograph of pancreatic acini. Permission from Dr. Richard A. Bowen Colorado State University. http://www.vivo.colostate. edu/hbooks/pathphys/ digestion/pancreas/histo_exo. html

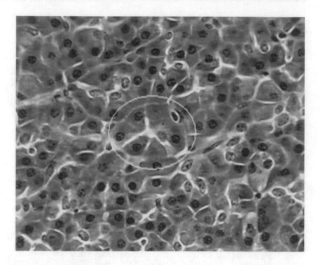

numerous digestive enzymes, including trypsinogen, chymotrypsinogen, procarboxypeptidases, ribonuclease, deoxyribonuclease, lipase, amylase, and proelastase, among others. Until stimulation, these constituents, and others are stored in zymogen granules of acinar cells.

6.2.3   Pancreatic Ducts

The pancreatic ductal system is similar to the parotid salivary gland ductal system with one major difference: There are *no striated ducts* in the pancreatic ductal system. The digestive enzymes from pancreatic acinar cells are initially secreted into a serous solution and flow via the ductal system where, ultimately, they are delivered into the *duodenum*.

Pancreatic ducts are classified into four types beginning with the terminal flattened cuboidal epithelium intercalated ducts that extend up into the lumen of the acinus to form a complex called centroacinar cells (Fig. 6.4). The exact role of centroacinar cells is still unknown, but they play an important role in maintaining the ionic content of the ductal lumen and keep it open for fluid flow. However, recent research suggests that these cells may also be involved with tissue homeostasis and regeneration.

The pancreatic fluid leaves the acinar region and enters the intercalated ducts (Fig. 6.5), which are composed of flattened cuboidal cells and secrete large amounts of HCO_3^- and Na^+ to the acinar fluid. Unlike the intercalated ducts of the salivary glands, they are without granules. The fluid then passes from the intercalated ducts to the intralobular ducts (Fig. 6.6). Intralobular ducts have a classical cuboidal epithelium and, as the name implies, are seen within lobules. The ducts are long, branched, and without secretory granules.

Fig. 6.4 The diagram shows the extension of intercalated cells into the pancreatic acinar lumen, which is not the case with the parotid acinar unit

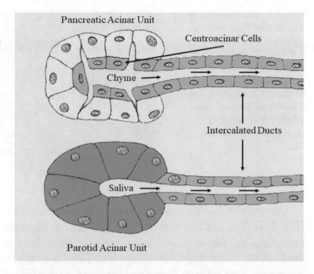

Fig. 6.5 The histology of tissues from the acinar unit (dashed arrow) and the intercalated ducts (solid arrow). Note the flattened cuboidal cells. Permission from Dr. Richard A. Bowen Colorado State University. http://www.vivo.colostate.edu/hbooks/pathphys/digestion/pancreas/histo_exo.html

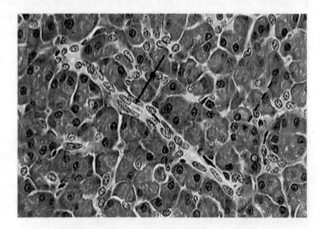

Fig. 6.6 The histology of an intercalated duct emptying into an intralobular duct. Note the cuboidal epithelium in the intralobular duct as indicated by the arrow in the slide. Permission from Dr. Richard A. Bowen Colorado State University. http://www.vivo.colostate.edu/hbooks/pathphys/digestion/pancreas/histo_exo.html

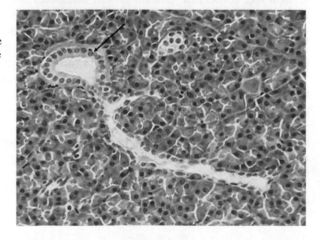

Fig. 6.7 The histology of an interlobular duct with its columnar epithelia as indicated by the arrow in the slide. Permission from Dr. Richard A. Bowen Colorado State University. http://www.vivo.colostate.edu/hbooks/pathphys/digestion/pancreas/histo_exo.html

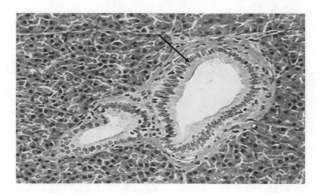

Interlobular ducts are found between lobules, within the connective tissue septae (Fig. 6.7). The smaller ducts have a cuboidal epithelium, while a columnar epithelium lines the larger ducts. Intralobular ducts deliver secretions from intralobular ducts to the *major pancreatic duct* also known as the *duct of Wirsung*. The major pancreatic duct exhibits a stratified columnar epithelial lining with the presence of an occasional goblet cell. The major pancreatic duct eventually penetrates the wall of the duodenum to deposit its secretions. The pancreatic duct joins the bile duct before entering the intestine.

6.3 Liver

Similar to the pancreas, the liver plays both endocrine and exocrine roles. Its endocrine role involves synthesizing and releasing plasma proteins, such as lipoproteins, prothrombin, fibrinogen, and albumins, into the bloodstream. The liver is also involved in detoxification; in lipid, carbohydrate, and protein metabolisms; and storage of iron, blood, glycogen, triglycerides, and vitamins. Its exocrine function is the production of *bile*.

Bile can be defined as a liquid that emulsifies and degrades fat into smaller molecules. The substance also aids in eliminating waste from the body. Bile contains water, *Bauhin's gland* salts, lecithin, cholesterol, fatty acids, bilirubin, and electrolytes.

Bile is produced by hepatocytes and is collected by bile canaliculi (Figs. 6.8 and 6.9). It drains into the hepatic duct, then into the cystic duct, and finally enters the gallbladder. A small space between hepatocytes and the endothelium of the sinusoids is called the *perisinusoidal space* or *space of Disse* (Fig. 6.8a). The endothelium of sinusoids enables proteins, nutrients, wastes, and plasma components from hepatic sinusoids to enter the space of Disse. Hepatocytes take up nutrients and transport wastes, such as bilirubin, into the bile. The central veins collect the exchanged blood from sinusoids and drain into the sublobular veins and then into the large hepatic veins.

The basic structure of the liver includes the classic lobule (Fig. 6.8b), the portal lobule, the liver acinus, and hepatocytes (Fig. 6.9).

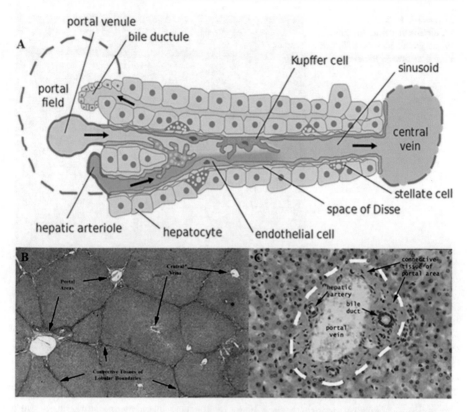

Fig. 6.8 Liver tissue anatomy (**a**), liver lobule histology (**b**), and the triad system (**c**). Permission from Dr. David G. King of Southern Illinois University

Fig. 6.9 The histology of hepatocytes. Note the numerous hepatocytes that exhibit polyploidy as indicated by the arrows in the slide

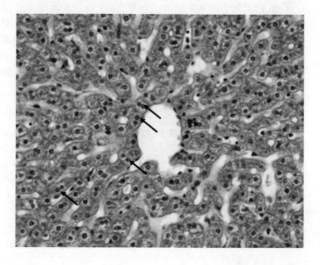

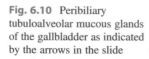

Fig. 6.10 Peribiliary tubuloalveolar mucous glands of the gallbladder as indicated by the arrows in the slide

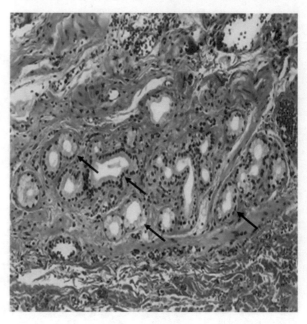

6.4   Gallbladder

The gallbladder is a reservoir for the storage of bile and is composed of surface epithelium, a lamina propria with a layer of loose smooth tunica muscularis, and perimuscular connective tissue. A unique feature of the gallbladder is its lack of a muscularis mucosae and submucosa.

The gallbladder contains peribiliary tubuloalveolar mucous glands that appear small and round. The glands are composed of cuboidal cells containing mucin and a basal nucleus. The glands are arranged in a lobular fashion (Fig. 6.10) and are separated by a fibrous stroma. The glands are found mainly in the neck of the gallbladder, as well as throughout the extrahepatic bile ducts.

Questions

1. The pancreas has striated ducts. Is this statement true or false?
    a. True
    b. False
2. Which of the following is a compound tubuloacinar serous organ?
    a. Pancreas
    b. Liver
    c. Gall bladder
3. Which of the following have a cellular structure similar to the salivary glands?

    a. Pancreas

    b. Liver

    c. Gallbladder

4. Both the salivary glands and the pancreas secrete amylase. Is this statement true or false?

    a. True

    b. False

5. Which of the following organs contain peribiliary tubuloalveolar mucous glands?

    a. Pancreas

    b. Liver

    c. Gallbladder

6. The space of Disse is found in which of the following organs:

    a. Pancreas

    b. Liver

    c. Gallbladder

Exocrine Glands of the Reproductive System

7

Abstract

The exocrine glands of the reproductive system are numerous, complex, and provide a wide variety of functions. The glands vary according to gender and in this chapter the information starts with a glandular description of women. The chapter details the mammary glands and the minor glands associated with its overall function and proceeds to describe the internal and exocrine glands of the female reproductive system. With the female reproductive system concluded, it then describes the male reproductive system. Similar to the female reproductive system, the chapter relates the major and minor glands of the male reproductive system. It will describe the major glands such as the prostate, semi vesicles, and bulbourethral glands. The minor glands are also discussed.

Learning Objectives

After reading the chapter, the reader should know the following concepts:

1. To educate the reader on the exocrine function of the mammary glands.
2. To educate the reader on the exocrine glands of the female reproductive system.
3. To educate the reader on the exocrine glands of the male reproductive system.

7.1 Introduction to the Female Reproductive System

The human female reproductive system contains two internal parts which are the uterus and the ovaries. The uterus is joined with the vagina at the cervix and meets the external organs at the vulva. The vulva includes the labia, clitoris, and urethra. In addition, the uterus is connected to the ovaries via the fallopian tubes. Amidst this complex system are numerous exocrine glands. The ensuing paragraphs will provide the histology, site, and function of these exocrine glands. The mammary glands are not genitalia, but are included in this section due to function during and after pregnancy.

7.1.1   Mammary Glands

Mammary gland tissues are compound exocrine glands composed of specialized glandular epithelium originating from the ectodermal germ layer. They are generally associated with the female reproductive system. The inner surface of the glands and the ducts that drain them are topologically continuous with the exterior of the body and they produce a secretion that eventually exits into the *extracorporeal* milieu. The glands primarily consist of parenchyma and a supporting connective tissue frame-work or stroma as illustrated in Fig. 7.1.

The tissues contain two epithelial cell types—ductal and acinar cells (*paren-chyma*)—along with myoepithelial cells, which contract to move fluid from the acinar lumen to the ducts. The *acinar cells* (Fig. 7.2) perform the special function of the organ-producing milk. The ductal tissue has 10–15 branching ducts (tube canals), which also carry glandular secretions. These ducts then merge into what is referred to as a primary duct, which then drains into the opening of the nipple. The

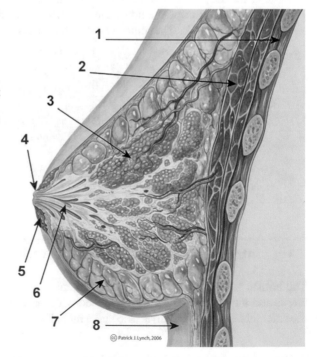

Fig. 7.1 The anatomy of the breast: (1) chest wall, (2) pectoralis muscle, (3) lobules, (4) nipple, (5) areola, (6) milk duct, (7) fatty tissue, and (8) skin. With permission from Patrick J. Lynch, medical illustrator. This file is licensed under the Creative Commons Attribution 3.0 Unported license

Fig. 7.2 A diagram of an acinar-ductal unit with cuboidal ductal cells (**a**), acinar cells (**b**), and myoepithelial cells (**c**)

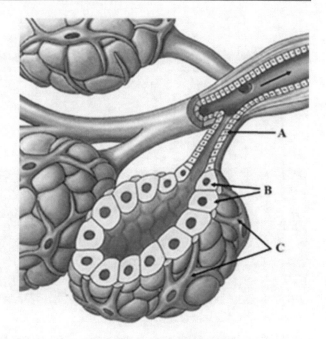

ductal epithelial cells (terminal ducts) adjacent to the acinar units are cuboidal. Surrounding the complex is the stroma, *which is the* tissue framework of the organ that provides support and material in which the parenchyma grows and is able to function.

The mode of secretion for the mammary glands is primarily apocrine. Materials such as lipids accumulate at the apical portion of the cell; when they reach a sufficient level, the cell "pinches-off," depositing the particulates into the lumen. Similar to pancreatic tissue, they are also without striated ducts. Unlike the intercalated and striated ducts of the salivary glands, sodium reabsorption does not occur in the intercalated ducts of the mammary glands; therefore, the resultant milk product is isotonic relative to serum. The ductal epithelium of the mammary principally functions as a conduit for the transport of milk to the nipple.

The most obvious yet striking difference between the other exocrine tissues is that mammary glands do not function through the course of the individual's life span. The development of the mammary glands to prominence occurs during puberty due to the influence of increased levels of ovarian steroids. It appears that estrogen promotes ductal development, while progesterone facilitates alveolar growth. During pregnancy, the placenta produces placental lactogen factor, which initiates further, the rapid growth of the tissue. Postpartum, *prolactin* becomes the most important hormone as it initiates and maintains milk production (Fig. 7.3).

Active lactation is achieved via a *neuroendocrine reflex* in which *oxytocin* is produced, resulting in the contraction of myoepithelial tissue that forces the milk to the nipple. Once the infant is weaned and nursing activities cease, the alveoli decrease in size and the tissues return to their prior prepregnancy state (Fig. 7.3).

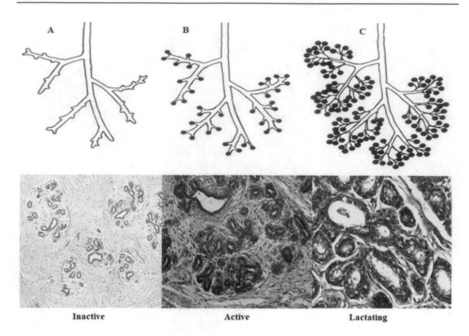

Fig. 7.3 The histology of mammary tissues at three distinct stages: the inactive state (**a**), the active state (**b**), and during lactation (**c**). Note a lack of acinar units in the inactive state (**a**). By contrast, the lactating state (**c**) has abundant acinar units with the presence of numerous lipid vesicles. Figure courtesy of the Creative Commons Attribution-Share Alike 4.0 International license

After menopause, a portion of the mammary lobules and ducts are obliterated and replaced with connective tissue or become calcified. No other exocrine gland functions in this capacity.

One important aspect is the areola of the breast, which has numerous types of exocrine glands including eccrine, apocrine, sebaceous, *apoapocrine*, and *glands of Montgomery*. Eccrine, apocrine, sebaceous, apoapocrine glands have been discussed and illustrated in Chap. 2.

Glands of Montgomery are a combination of sebaceous glands. They can release a small amount of breast milk, but they mostly produce an oily substance that lubricates and cleans the areola and nipple. The oil has antimicrobial constituents that prevent infection and tissue cracking (Fig. 7.4).

It is worth noting that apocrine sweat glands on the areola secrete pheromones that promote a mother–infant bond during lactation. The scents produced by the mother's pheromones are received by the vomeronasal organs of the infant; this phenomenon assists the child in identifying its mother for nursing.

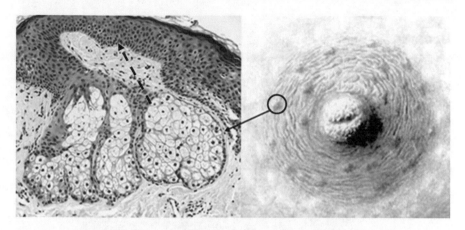

Fig. 7.4 The appearance of the areola (right) and the histology of glands of Montgomery as shown by the solid arrow. The dashed arrow points to the stratified squamous epithelium layer

7.2 Female Reproduction System

The female reproductive system is composed of internal sex organs and external genitalia. The internal reproductive organs are cradled within the pelvis, while the external *genitalia* are situated in the anterior part of the *perineum* or the *vulva*.

The internal reproductive system is composed of the ovaries, uterine tubes, uterus, and vagina, while the external genitalia are composed of the mons pubis, labia majora, labia minora, clitoris, vestibule, vaginal opening, hymen, and the urethral orifice (Fig. 7.5).

7.2.1 Exocrine Glands of the External Reproductive System

7.2.1.1 Greater Vestibular Glands

The *greater vestibular glands*, also called *Bartholin glands*, are two pea-sized compound alveolar glands located slightly posterior and to the left and right of the opening of the *vagina* (Fig. 7.6). The glands secrete mucus for microbial protection and possibly to lubricate the vagina. They are homologous to *bulbourethral glands* in men. The greater vestibular glands are composed of several types of epithelium. The gland is composed of acinar units that are lined with mucinous columnar epithelium (Fig. 7.7). These integrate with ducts that are composed of transitional epithelium, which eventually becomes squamous epithelium at the vaginal ostia.

7.2.1.2 Lesser Vestibular Glands (Skene's Glands)

Skene's glands, also known as the lesser vestibular glands or paraurethral glands, are mucinous glands on the anterior wall of the vagina, next to the lower end of the urethra (Fig. 7.8). The glands secrete a fluid that helps lubricate the urethral opening

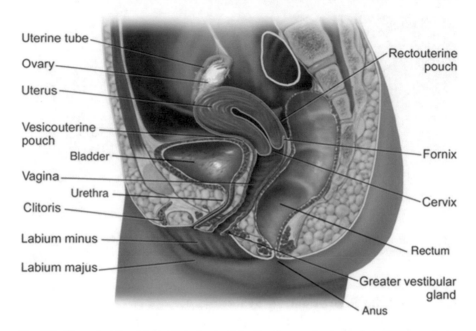

Fig. 7.5 The external and internal anatomy of the female reproductive system. With permission from Blausen.com staff (2014). "Medical gallery of Blausen Medical 2014". *WikiJournal of Medicine* 1 (2). DOI:10.15347/wjm/2014.010. ISSN 2002-4436

during intercourse. In addition, they are surrounded by tissue that swells with blood during sexual arousal.

7.2.2   Exocrine Glands of the Internal Reproductive System

7.2.2.1 Vagina
The vagina *does not* contain any exocrine glands within its walls and relies upon the cervical glands from the cervix for lubrication.

7.2.2.2 Cervix
Naboth's or cervical glands, located on the cervix, are highly branched and produce mucous secretions (Fig. 7.9). In recent studies, the cervical glands have been shown not to be "true" glands, but rather deep, cleft-like enfolding of the surface epithelium into the underlying stroma. The columnar cells (Fig. 7.9) of these glands have basal round or oval nuclei. The cytoplasm is finely granular and contains an abundance of mucin.

7.2.2.3 Oviduct
The epithelium of the oviduct comprises two distinct cell types: ciliated cells that serve to move the ovum toward the uterus and secretory peg cells. While ciliated

Fig. 7.6 Glands of the
external reproductive system

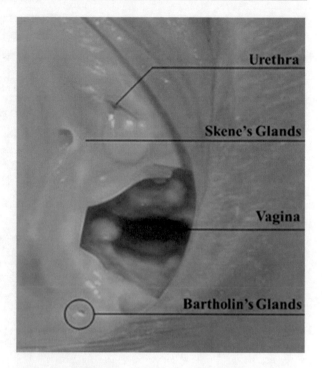

Fig. 7.7 The histology of the
greater vestibular glands
(Bartholin's gland), which are
composed of acinar units that
are lined with mucinous
columnar epithelium. Arrows
within the slide indicate
mucinous acini

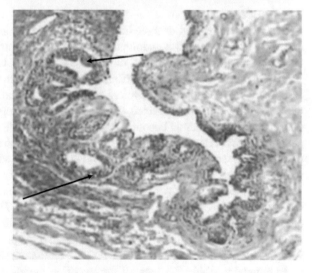

columnar cells move the ovum toward the uterus, Peg cells release a secretion that
lubricates the fallopian tube and provides nourishment and protection to the traveling
ovum (Fig. 7.10).

Fig. 7.8 The histology of
Skene's glands. The glands
surround the urethra with
stratified epithelium, ranging
from squamous to columnar
types

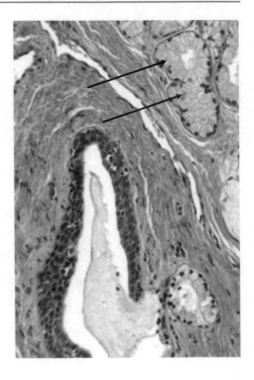

7.2.2.4 Uterus

The *uterus* is composed of three histological layers: the *endometrium*, *myometrium*, and the *perimetrium*. The endometrium is lined with simple columnar epithelium and contains numerous branched tubular glands (Fig. 7.11).

The uterine glands and the surrounding endometrium are affected by hormones associated with the menstrual cycle. During the proliferative phase (Fig. 7.12a), the uterine glands appear elongated due to estrogen secretion. During the secretory phase, the uterine glands become very coiled with wide lumina (Fig. 7.12b). This change corresponds with an increase in blood flow to spiral arteries due to increased progesterone secretion. The uterine glands are gorged with glycogen and suffused with stem cells.

During the menstrual phase (Fig. 7.12c), progesterone secretion decreases. This results in decreased blood flow to the spiral arteries. The functional layer of the uterus containing the glands becomes necrotic and eventually sloughs off during the menstrual phase of the cycle. In Fig. 7.12c, the lamina propria is breaking down and is replaced by blood. In addition, the surface epithelium is no longer present.

In summary, Table 7.1 lists all the exocrine glands associated with the female reproductive system.

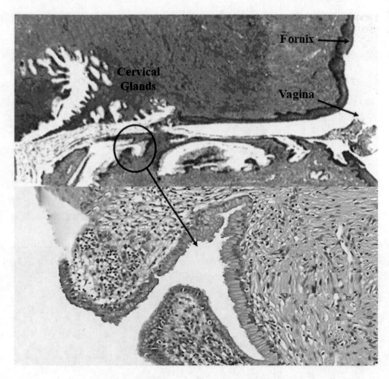

Fig. 7.9 The histology of the cervix showing the cervical glands. The branched cervical glands lined by tall columnar mucus-secreting cells are found in the underlying collagenous stroma

Fig. 7.10 The histology of
the oviduct with its two types
of cells: ciliated columnar
cells (**a**) and peg cells (**b**)

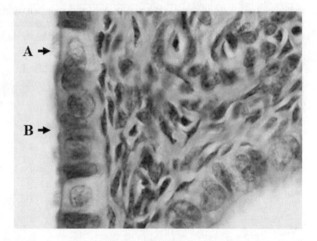

Fig. 7.11 The histology of
uterine glands. Arrows within
the slide point to the uterine
glands

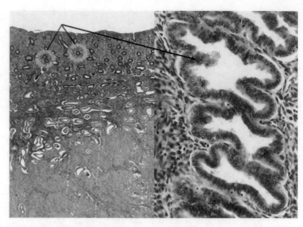

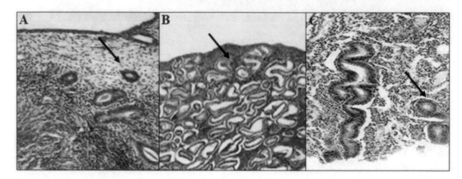

Fig. 7.12 The histological changes in the endometrium during the menstrual cycle: the
proliferative phase (**a**), the secretory phase (**b**), and the menstrual phase (**c**). The arrows indicate
uterine glands. Note how the uterine glands decrease markedly during the menstrual phase (**c**)

Table 7.1 Female reproductive system and its exocrine glands

Mammary glands			
Mammary gland	Breast	Milk	Compound-tubular/alveolar
Montgomery's glands	Breast	Sebum	Branched tubular
External glands			
Bartholin's glands	Vaginal ostia	Mucus	Tubuloalveolar
Integumental glands	Labia of the vulva	Perspiration	Coiled tubular
Skene's glands	Urethral ostia	Mucus	–
Internal glands			
Cervical glands	Cervix	Mucus	Simple branched
Peg cells	Fallopian tube	–	Nutritional fluid for the ovum
Endometrial glands	Endometrial	Mucus	–
Uterine glands	Uterus	Mucus	Tubular

7.3   Male Reproductive System

7.3.1   The Testes

The male reproductive system (Fig. 7.13) is a composite of varying tissues whose purpose is the production of spermatozoa, which are transported in a fluid called semen. The system is solely for procreation. From an exocrine perspective, the testes are made up of numerous compartments called lobules. Each lobule encompasses coiled seminiferous tubules, where the spermatozoa are made. After the sperm mature, the seminiferous tubules discharge the sperm into a series of ducts where they pass to the urethra. It is worth noting that spermatozoa are the *only cells* that have a flagellum in the human body.

The endocrine function of the testes is concerned with the production of testicular androgens such as testosterone and dihydrotestosterone (from the Leydig cells), and Müllerian inhibiting factor (from the Sertoli cells) are well documented in numerous textbooks and journals and are not discussed in this chapter; however, the exocrine glands involved with semen production (accessory glands) are described in the ensuing paragraphs, starting with the prostate gland.

7.3.2   Prostate Gland

The prostate is a tubuloalveolar exocrine gland that, in part, produces the fluid that carries and nourishes sperm during ejaculation. The prostate is a spongy,

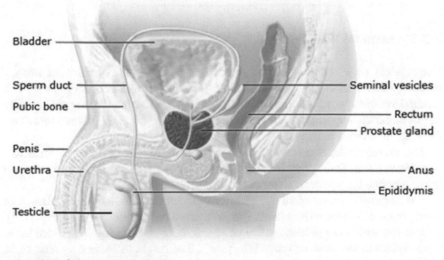

Location of the prostate gland

Fig. 7.13 The anatomy of the male reproductive system

heart-shaped gland and weighs about 30 g. It is located directly below the bladder and above the muscles of the pelvic floor. The ducts in the prostate gland flow into the urethra, which passes through the prostate (Fig. 7.13).

The prostate gland is surrounded by a capsule of connective tissue containing many smooth muscle fibers and elastic connective tissue. There are also many smooth muscle cells inside the prostate. During ejaculation, these muscle cells contract and forcefully press the fluid stored in the prostate into the urethra. This action causes the prostate fluid, spermatozoa, and fluid from other glands to combine to form semen, which is then released outside of the body.

The tissue of the prostate gland is divided into three different layers, which encircle the urethra. The tissue layers are as follows:

1. The *transition zone* is the smallest part of the prostate and is found on the innermost portion of the gland. It surrounds the urethra between the bladder and the upper third of the urethra.
2. The *central zone* surrounds the transition zone and makes up about one-quarter of the prostate's total mass. This is where the duct common to the prostate, the seminal duct, and the seminal vesicles are found.
3. The *peripheral zone* represents the main part of the prostate gland and is approximately 70% of the total prostate tissue mass.

The transition zone tissue is subject to aging and tends to undergo a benign growth termed *benign prostatic hyperplasia* (BPH). If this tissue presses against the bladder and the urethra, it can lead to urinary hesitancy. This is a common problem among older men. Contrary to BPH, malignant tumors in the prostate mostly develop in the peripheral zone.

7.3.3   Seminal Vesicles

The *seminal vesicles* are a pair of glands that are positioned below the urinary bladder. Each vesicle consists of a single tube folded and coiled on itself. The seminal vesicle spans 5–7 cm, although its full unfolded length is approximately 8–11 cm. Most of its length is curled inside the gland's structure. In some instances, there are diverticula in its wall.

The excretory ducts unite with their corresponding vas deferens to form two ejaculatory ducts. These ducts immediately pass through the prostate gland and open separately into the *verumontanum* of the prostatic urethra.

Histologically, the seminal vesicles have a mucosa, consisting of a lining of interspersed columnar cells, a lamina propria, and a thick muscular wall (Fig. 7.14).

The epithelium is pseudostratified columnar in character (Fig. 7.14), similar to other tissues in the male reproductive system. The height of these columnar cells, and therefore their activity, is dependent upon blood testosterone levels. The lamina propria contains small underlying blood vessels and lymphatics. Together with the epithelia, this is called the mucosa; it is arranged into convoluted folds, increasing

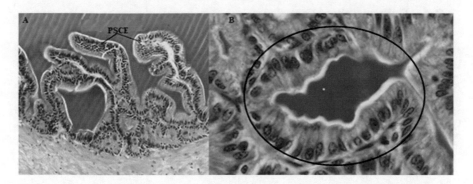

Fig. 7.14 The histology of the seminal vesicles. Panel (**a**) is a low-power micrograph that shows the muscularis and pseudostratified columnar epithelium (PSCE). Panel (**b**) illustrates the circled, dark-colored lumen surrounded by the pseudostratified columnar epithelium

Table 7.2 Components of semen in mL/ejaculate

Site	Volume	Composition
Testis and epididymis	0.15 (5%)	Spermatozoa (80 million/mL)
Seminal vesicles	1.5–2 (50–65%)	Fructose phosphorylcholine, ergothioneine, ascorbic acid, flavins, prostaglandins, bicarbonate
Prostate	0.6–0.9 (20–30%)	Spermine, citric acid, cholesterol, phospholipids, fibrinolysin, fibrinogenase, zinc acid, phosphatase, prostate-specific antigen
Bulbourethral gland	<0.15 (5%)	Mucus

the overall surface area. A muscular layer, consisting of an inner circular and outer longitudinal layer of smooth muscle, can also be found.

The seminal vesicle glands secrete a large portion (50–65%) of the ejaculate, which is called semen. The seminal vesicle secretion is mildly alkaline and yellowish in color. The yellowish color is attributed to lipofuscin granules from epithelial cells. The alkalinity of semen helps neutralize the acidity of the vaginal tract, prolonging the lifespan of sperm. As shown in Table 7.2, seminal vesicle secretions contain proteins, enzymes, fructose, mucus, vitamin C, flavins, phosphatidylcholine, and prostaglandins.

The development and maintenance of the seminal vesicles, as well as their secretions, are androgen-dependent. The seminal vesicles contain 5α-reductase, which metabolizes testosterone into its more potent metabolite dihydrotestosterone. The seminal vesicles are also regulated by the ligand of this receptor, luteinizing hormone.

7.3.4   Bulbourethral Gland

The bulbourethral glands (*Cowper's glands*) are a pair of exocrine glands in the male reproductive system. They are roughly the size of peas and are located inferior to the prostate gland and lateral to the urethra in the urogenital diaphragm. They are compound tubuloalveolar glands that secrete a mucus-like substance containing galactose, galactosamine, galacturonic acid, sialic acid, and methylpentose. The function of the secretion is to lubricate the urethra and neutralize acidic urine. The bulbourethral glands are only found in the male body and play an important role in the protection of sperm during ejaculation (Fig. 7.15).

7.3.5   Miscellaneous Male Reproductive Exocrine Glands

7.3.5.1   Preputial Gland (Tyson's Glands)
The preputial glands were discovered by Edward Tyson and later described by William Cowper in 1694. They are described as modified sebaceous glands. These glands are located around the corona and inner surface of the prepuce of the human penis (Fig. 7.16). They are holocrine, branched tubuloalveolar structures with acini that open into lateral ducts. Their secretions contribute to the composition of *smegma*.

Smegma is composed of 13% proteins and 27% fats. These constituents are the result of decomposing epithelial cells and their debris. Smegma has a smooth, moist texture, and is rich in squalene, a sterol precursor. It also contains prostatic and seminal secretions, desquamated epithelial cells, and the mucin content of the periurethral glands. Cathepsin B, lysozymes, chymotrypsin, neutrophil elastase, and cytokines are also present in smegma.

7.3.5.2   Periurethral (Littré Glands)
The periurethral glands (Littre glands) branch off the wall of the urethra among men. The glands are mucinous and are primarily present in the section of the urethra that

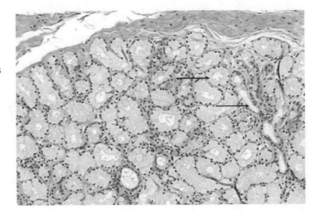

Fig. 7.15 The histology of bulbourethral glands. The glands are mucinous in appearance and have simple columnar cells lining the ducts

Fig. 7.16 The histology of
preputial glands, which are
sebaceous. Additionally, they
are located around the corona
and the inner surface of the
prepuce of the human penis.
Circled within the slide is an
example of one of the glands
surrounded by squamous
epithelium

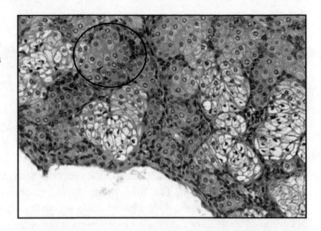

Fig. 7.17 The histology of
the urethra (U) and
periurethral glands (P)

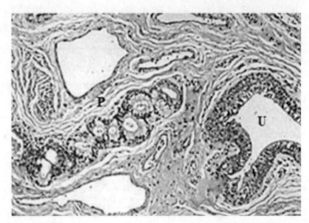

courses through the penis (Fig. 7.17). The periurethral glands secrete
glycosaminoglycans that protect the urethral epithelium against acidic urine.

7.3.5.3 Scrotum

The scrotum is an external pouch housing the testes whose function is to maintain the
proper temperature for spermatozoa. The skin is thin, corrugated, and pigmented. It
includes keratinized squamous epithelium with skin adnexa, dermis, and scattered
adipocytes but no subcutaneous tissue. The pouch is divided in half by a midline
cutaneous raphe, which continues to the inferior penile surface and along the
perineum to the anus. As described in Chap. 2, the skin of the scrotum has hair,
sebaceous glands, and amine precursor uptake decarboxylase (APUD) cells. Scent
glands are also present and produce pheromones.

7.3.6   Semen

Semen is the product of the male reproductive system. In healthy men, it is a white opalescent fluid with a pH between 7.35 and 7.50. Approximately 3 mL is produced by ejaculation, with a specific gravity of 1.028. Table 7.2 lists the composition of semen according to what each tissue contributes.

Semen is a composite of fluids from several secretory tissues. It is composed of sperm cells from the testicles, fluid from the seminal vesicle, and secretions released by the bulbourethral gland and the prostate fluid. All of these fluids are blended together in the urethra.

Prostatic secretion is important for proper spermatozoa functioning and, therefore, male fertility. The thin, milky liquid contains many enzymes such as *prostate-specific antigen*. This glycoprotein enzyme makes the semen thinner. The hormone-like substance *spermine* mostly ensures sperm cell motility. In addition, in the prostate the hormone testosterone is transformed to a biologically active form designated as dihydrotestosterone.

Questions

1. The Skene's cells produce a serous secretion. Is this statement true or false?
   a. True
   b. False
2. Which of the following is a compound tubuloacinar gland?
   a. Mammary gland
   b. Testes
   c. Bulbourethral gland
   d. Uterine glands
3. The organs of reproduction can be considered both exocrine and endocrine. Is this statement true or false?
   a. True
   b. False
4. All of the following glands are mucinous except one? Identify the non-mucinous gland.
   a. Mammary gland
   b. Endometrial glands
   c. Bulbourethral gland
   d. Uterine glands
5. The internal glands of the female reproductive system are primarily mucinous. Is this statement true or false?
   a. True
   b. False
6. Which one of the following compounds is produced primarily by the seminal vesicles?
   a. Zinc acid
   b. Ergothioneine

     c. Cholesterol

     d. Spermine

7. The non-ciliated secretory cells, also known as peg cells, release a secretion that lubricates the oviduct and provides nourishment and protection to the traveling ovum. Is this statement true or false?

     a. True

     b. False

8. The Bartholin's glands are two pea-sized compound alveolar glands located slightly posterior and to the left and right of the opening of the vagina. Is this statement true or false?

     a. True

     b. False

Suggested Reading

Bailey, Frederick, Wood, Richard, Enders, Allen, and Kelly, Douglas. 1984. "The Female Reproductive System". In *Bailey's Textbook of Microscopic Anatomy*, 18th ed., edited by Douglas Kelly, Richard Woods, and Allen Enders, 769–775. Baltimore: Williams & Wilkins.

Boschat, Corina, Pélofi, Coryse, Randin, Olivier, , Daniele, Lüscher, Christian, Broillet, Marie-Christine and Rodriguez, Ivan. 2005. "Pheromone Detection Mediated by a V1r Vomeronasal Receptor". *Nature Neuroscience*, Vol. 5: 1261–1262. https://doi.org/10.1038/nn978

D'Aniello, Biagio, Semin, Gün, Scandurra, Anna, and Pinelli, Claudia. 2017. "The Vomeronasal Organ: A Neglected Organ". *Frontiers in Neuroanatomy*. Vol. 11: 70. https://doi.org/10.3389/fnana.2017.00070.

Filant, Justyna, and Spencer, Thomas. 2014. "Uterine Glands: Biological Roles in Conceptus Implantation, Uterine Receptivity, and Decidualization". *International Journal of Developmental Biology*, Vol. 58: 107–116.

McManaman, James, Reyland, Mary, and Thrower, Edwin. 2006. "Secretion and Fluid Transport Mechanisms in the Mammary Gland: Comparisons with the Exocrine Pancreas and the Salivary Gland". *Journal of Mammary Gland Biology and Neoplasia*, Vol. 11: 249–268.

Nicholson, Bubba. 2011. *Exocrinology The Science of Love: Human Pheromones in Criminology, Psychiatry, and Medicine*. Tampa: Nicholson Science.

Exocrine Glands of the Urinary System

8

Abstract

The main question concerning the kidney is whether it is an endocrine or exocrine gland. The answer is, it is both. The ensuing pages will explain why the kidney is also an exocrine organ.

Learning Objectives

After reading the chapter, the reader should know the following concepts:

1. The reader should be able to explain the exocrine function of the renal system.

8.1   Introduction

The kidney is an endocrine gland because it produces hormones such as *renin*, *angiotensin*, components of the *kinin–kallikrein system, erythropoietin*, and *prostaglandins*. These hormones act to influence blood pressure, sodium and water excretion, red blood cell production, calcium homeostasis, and the immune system. They are endocrine because they have additional ductless cell groups that produce the hormone erythropoietin and they have roles in producing two other hormones, angiotensin, and calcitriol.

The kidney is also an exocrine gland because it has ducts (the *renal tubules*) that excrete nitrogenous wastes in the urine. Taken together, this fits the definition of exocrine function. Consequently, this chapter will examine *only* the exocrine function of the urinary system.

8.2   The Exocrine Kidney

8.2.1   The Glomerulus

The kidney is a compound tubular gland located in the *retroperitoneal* portion of the body. As a homeostatic organ, it has an important role in regulating the composition of extracellular fluids as well as eliminating urea and other nitrogenous waste material.

The primary unit of the aforementioned functions is the *nephron*, which is the tubular structure composed of *Bowman's capsule* and the *glomerulus* (Fig. 8.1). Together they form the *renal* or *Malpighian corpuscle*.

Bowman's capsule consists of two *epithelial layers*. The first layer is the *visceral layer*, which covers the glomerulus, and the second, *parietal* or outer layer is continuous with the visceral layer at the edge of the opening to the renal corpuscle. The visceral layer contains *podocytes*, which are unique because their nucleus and cytoplasm protrude into the *capsular space*. The remaining processes extend from the cell body to the basal lamina. These processes are called *pedicels* (Fig. 8.2).

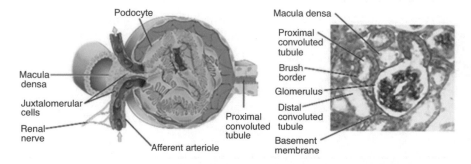

Fig. 8.1   A diagram (left) and the histology (right) of the glomerulus and surrounding structures

Fig. 8.2 As indicated by arrows, the scanning electron micrograph shows the interdigitation of podocyte processes

Fig. 8.3 The electron micrograph shows a glomerular capillary wall

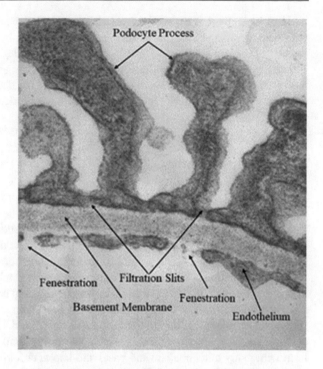

Pedicels interdigitate with each other to produce a layer of slits between them (Fig. 8.2). These slits are approximately 0.02–0.04 μm in width and are covered by an electron-dense membrane called the *slit diaphragm* that is 5.0 nm thick (Fig. 8.3). The slit membrane is filamentous and has an arrangement of pores that are 5–15 nm. Small molecules such as urea pass through the single *glomerular basement membrane* between the endothelium and podocytes through the slit membrane into the urinary space of Bowman's capsule (Fig. 8.3).

The parietal layer is far less complicated than the visceral layer. The parietal layer initiates where the visceral layer manifests at the vascular pole. It is composed of simple epithelium resting on the basal lamina enveloped by connective tissue. In addition, the epithelial cells are uniform in thickness and eventually become cuboidal in shape.

8.2.2   Proximal Convoluted Tubule

The *proximal convoluted tubule* is situated between Bowman's capsule and the *loop of Henle* and is essential for the resorption of sugar, Na^+, Cl^-, and water from the glomerular filtrate. It also regulates the pH of the filtrate by exchanging H^+ in the interstitium for HCO_3^- in the filtrate. In addition, the proximal convoluted tubule is responsible for the reabsorption of most of the filtrate (85%), which includes glucose, amino acids, electrolytes, and water.

Histologically, the cells of the proximal convoluted tubule are simple cuboidal epithelial. The large cells, however, are abundant with microvilli that are approximately 1.0 μm in length and form a brush border at the apical surface (Fig. 8.1). Taken together, the resultant diameter of the tubule is small. The basal membrane is folded to increase the surface area. The cells are also characterized by their large appearance, numerous mitochondria, and indistinct borders.

8.2.3 Loop of Henle

The "U-Shaped" loop of Henle is a continuation of the proximal convoluted tubules with five basic components: the thick descending limb, the thin descending limb, the thin hairpin, the thin ascending limb, and the thick ascending limb.

The thin segment of the loop of Henle (Fig. 8.3) has a diameter of approximately 12 μm and consists of low cuboidal to flattened squamous cell epithelium that has nuclei protruding into the lumen. The thin descending limb has low permeability to ions and urea but is highly permeable to water. The loop has a sharp bend called the hairpin loop, which forms the transition from descending to the ascending loop.

The thick ascending limb of the loop of Henle (Fig. 8.4) is lined by low cuboidal cells, which lack a brush border. The thin ascending limb is a low cuboidal epithelium with many mitochondria and basal and lateral enfolding of the cell membrane. This section of the loop of Henle is impermeable to water and actively pumps chloride out to the interstitium with sodium passively following chloride.

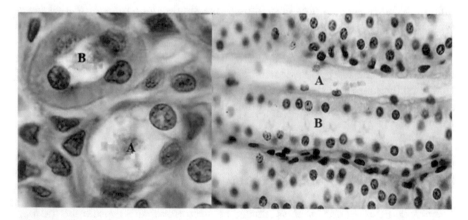

Fig. 8.4 The cross-sectional (left) and horizontal (right) histology of the thick and thin segments of the loop of Henle. **a**, the thin segment; **b**, the thick segment

8.2.4   Distal Convoluted Tubule

The *distal convoluted tubule* is a tortuous structure situated between the loop of Henle and the collecting tubule. It is the end portion of the nephron and is lined with simple cuboidal epithelium. The main function of the distal convoluted tubule epithelium is the reabsorption or excretion of HCO_3^- and H^+ to maintain blood pH. Unlike the proximal convoluted tubule, the epithelia are smaller and without the brush appearance of the numerous microvilli. The cells of the distal convoluted tubule are also less acidophilic because they have fewer mitochondria. The cells do possess membrane invaginations and lateral extensions. Similar to the proximal convoluted tubule, lateral cell margins are difficult to distinguish. Figure 8.5 illustrates the cell structure.

8.2.5   Collecting Ducts

Urine passes from distal convoluted tubules to the collecting ducts, which originate as small diameter tubules (40 μm) consisting of cuboidal epithelium. As the tubules progress toward the papillae, the cells become taller until they transform into columnar epithelium. At this point, the tubules are 200 μm in diameter. Figure 8.6 illustrates the histological appearance of the collecting ducts.

Fig. 8.5 The presence of cuboidal cells of the distal convoluted collection duct as indicated by the arrows

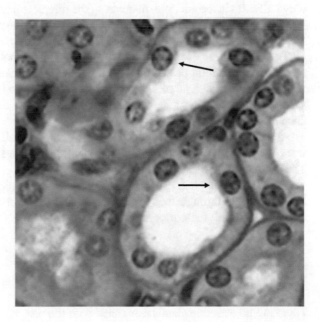

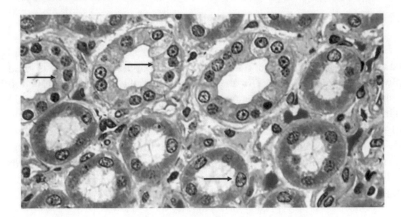

Fig. 8.6 The histology of collecting ducts. Arrows indicate cuboidal epithelium

8.2.6 Synopsis of Urine Formation

The renal system's primary function is to regulate blood volume, osmolality, and waste removal via urine. Urine formation occurs during three processes: (1) *filtration*, (2) *reabsorption*, and (3) *secretion*.

During filtration, blood enters the afferent arteriole and flows into the glomerulus, where filterable blood components, such as water and nitrogenous waste, will move toward the inside of the glomerulus. With respect to the nonfilterable components, they will exit via the arteriole. In contrast, the filterable particulates accumulate in the glomerulus forming a filtrate. Approximately 20% of the total blood pumped by the heart will enter the kidney to undergo filtration.

The next phase is reabsorption. At this stage, molecules and ions will be reabsorbed. The fluid passes through the renal components that include the proximal/distal convoluted tubules, the loop of Henle, and the collecting duct. It is at this stage of fluid passage that water and ions are removed as the fluid osmolality changes.

At the secretion phase, substances such as H^+, creatinine, and drugs are removed from the blood through the peritubular capillary network into the collecting duct. The product of all these processes is urine.

At the secretion phase, substances such as H^+, creatinine, and drugs are removed from the blood through the peritubular capillary network into the collecting duct. The product of all these processes is urine.

8.3   Ureters

There are no exocrine glands associated with the ureters.

8.4   Bladder

There are no exocrine glands associated with the bladder.

8.5   Urethra

The urethra is different between men and women; consequently, it is described in Chap. 7.

8.6   Urine

Urine is the excreted by-product of metabolism. It contains ammonia, uric acid, urea, sIgA, and creatinine cleared from the bloodstream. The normal individual produces approximately 1.4 L/day (range 0.5–2.5 L/day) with an average pH of 6.2 (pH range 5.5–7). It is a clear yellow liquid with a specific gravity of 1.003–1.035 and contains an assortment of inorganic salts, including proteins, hormones, and a wide range of metabolites. Collectively, the total solids in urine are approximately 59 g/day; this amount depends on dietary intake and varies according to health status and pharmacological usage. Similar to blood, urine is a primary diagnostic medium that has been used for decades.

Questions

1. There are numerous exocrine glands associated with the ureters and the bladder. Is this statement true or false?
   a. True
   b. False
2. Urine contains which of the following compounds?
   a. Urea
   b. Creatinine
   c. Ammonia
   d. Secretory IgA
   e. a, b, c only
   f. All of the above
3. The kidney is a compound tubular gland. Is this statement true or false?
   a. True
   b. False
4. Describe the difference between secretion and excretion.
5. The distal convoluted tubule has smaller epithelia and does not have microvilli. Is this statement true or false?
   a. True
   b. False

6. Explain why the kidney is also an exocrine gland.
7. Explain the function of the Loop of Henle.

Suggested Reading

Kelly, Douglas, Wood, Richard, and Enders, Allen. 1984. "The Urinary System". In *Bailey's Textbook of Microscopic Anatomy*, 18th ed., edited by Douglas Bailey, RL Woods, and AC Enders, 150–155. Baltimore: Williams & Wilkins.

Junqueira, Luis, and Carneiro, Jose. 1983. "Urinary System Tract". In *Basic Histology*, 4th ed., edited by Luis Junqueira and Jose Carneiro, 394–416. Los Altos: Lange Medical Publications.

Pawlina, Wojciech. 2016. "Urinary System I". In *Histology: A Text and Atlas with Correlated Cell and Molecular Biology*, 7th ed., edited by Wojciech Pawlina, 554–567. Philadelphia: Wolters Kluwer Health.

Exocrine Glands of the Sensory Organs

9

Abstract

This chapter identifies and describes the exocrine glands of the ear and the eye. The ear is simple with respect to the presence of exocrine glands having only three glands to understand with regard to type, site, size, and function. In contrast, the eye is complex having numerous exocrine glands that formulate tears. This chapter will elucidate the importance of these and how they unite to form the protective substance called tears.

Learning Objectives

After reading the chapter, the reader should know the following concepts:

1. The reader should be able to identify the exocrine glands of the ear.
2. The reader should be able to identify the exocrine glands of the eye.

9.1   The Ear

The external ear has a framework of elastic cartilage, which is continuous with that of the pinna and varies greatly in shape and size among individuals (Fig. 9.1). The internal elastic cartilage of the ear is covered by a layer of skin on both sides of the structure. The skin contains sweat and sebaceous glands in the dermis of the tissue. There are three types of glands in the external ear: sebaceous, eccrine sweat (merocrine sudoriferous), and ceruminous (wax) glands.

The sebaceous glands are large and numerous at the tragus and orifice of the auditory canal. They become smaller and less numerous toward the inner portion of the cartilaginous canal and eventually disappear in the innermost portion of the tube. The eccrine sweat glands are present in the concha at the level of the tragus and dispel

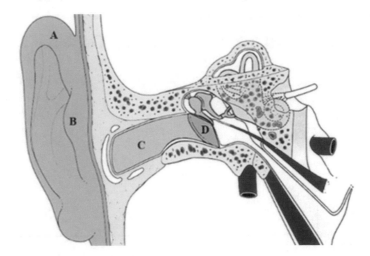

Fig. 9.1 The diagram shows the external ear (**a**), the tragus (**b**), the auditory meatus (**c**), and the tympanic membrane (**d**)

Fig. 9.2 The histology of the auditory meatus, including the auditory canal (**a**), sebaceous glands (**b**), a gland duct (**c**), and ceruminous glands (**d**)

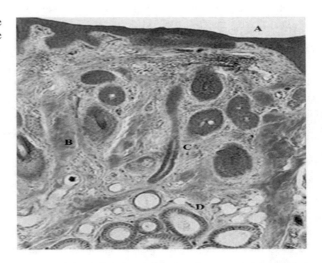

at the orifice of the canal. The eccrine sweat glands are replaced by large, alveolar, ceruminous glands, which are a special variety of coiled tubular apocrine sweat glands.

While the auditory meatus contains some ceruminous and sebaceous glands, it has no eccrine sweat glands. The ceruminous glands lie, in aggregates of six or more, deep in the connective tissue layer of the ear canal, often resting upon the perichondrium of the cartilaginous tube. These glands are like those of apocrine glands elsewhere in the body. They are tall and have a comparatively narrow lumen (Fig. 9.2). The secretory cells of the ceruminous glands rest on a layer of myoepithelial cells, which in turn rest on a thick, hyalinized basement membrane.

Thus, as in the cases of eccrine sweat glands, these ceruminous glands possess the power of contraction. The ducts of these glands usually open directly onto the skin surface; they occasionally open into sebaceous ducts. Detailed information can be found in Chap. 2.

9.2   The Eye

The eye is a complex photosensitive organ that allows an accurate analysis of light intensity and color from objects. The components of the eye are diagrammed in Fig. 9.3.

The eye has numerous exocrine glands that vary in type, size, secretion, and function. In addition to the lacrimal gland, there are seven additional exocrine glands: glands of *Henle, Krause, Meibomian, Manz, Moll, Wolfring (Ciaccio),* and *Zeis* (Table 9.1). There are also numerous goblet cells.

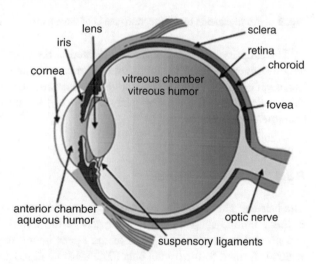

Fig. 9.3 An illustration of the anatomy of the eye

Table 9.1 A summary of the exocrine glands of the eye

Gland	Type	Secretion	Structure
Glands of Wolfring	Merocrine	Tears	Tubular
Glands of Zeis	Holocrine	Sebum	Tubular
Goblet cells	Merocrine	Mucin	Unicellular
Henle's glands	Apocrine	Mucin/tears	Tubular
Krause's glands	Merocrine	Tears	Tubular
Lacrimal glands	Merocrine	Tears	Tubuloacinar
Manz glands	Merocrine	Mucus	Tubular
Meibomian (tarsal) gland	Holocrine	Lipids/tears	Tubuloacinar
Moll's gland	Apocrine	Sebum	Tubular

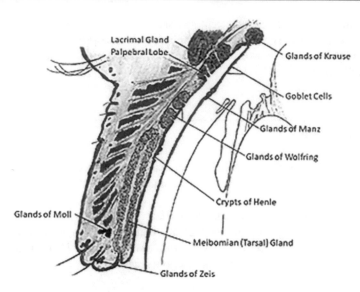

Fig. 9.4 An illustration of the exocrine glands of the eyelid

The exocrine glands are primarily located in the palpebral (eyelids) of both eyes and are generally more predominant in numbers in the upper eyelid compared with the lower eyelid (Fig. 9.4). Taken together, the exocrine glands produce a composite fluid called tears. The ensuing paragraphs detail the histology and function of the aforementioned glands.

9.2.1 Lacrimal Glands

The lacrimal glands are the major exocrine glands of the eye. Each eye contains one of these almond-shaped glands, which secrete the serous layer of the tear film. They are situated in the *lacrimal fossa* of the upper lateral region of each *ocular orbit*, which is formed by the frontal bone. The lacrimal gland produces tears that flow into secretory ducts and eventually empty onto the surface of the eye. Blinking assists in covering the entire surface of the *cornea* with tears that continue to flow and are eventually finally collected by the *lacrimal puncta*. The lacrimal puncta are connected to the *lacrimal canaliculus* that empties into the *lacrimal sac*. From the lacrimal sac, the tears flow through the *nasolacrimal duct* into the nasal cavity below the *inferior turbinate*.

Histologically, the lacrimal gland (Fig. 9.5) is a compound tubuloacinar gland. The gland consists of many lobules separated by connective tissue. Within each lobule, there are numerous acini. The acini are composed of large serous cells that produce a watery secretion. In addition, serous cells contain numerous lightly stained secretory granules and are surrounded by well-developed myoepithelial cells long with a sparse, vascular stroma (Fig. 9.5). Each acinus consists of a grape-like mass of

Fig. 9.5 The histology of a lacrimal gland. Note the acinar units in the middle of the image (circled on slide) and ductal structures near the bottom of the slide (within a dashed circle on slide)

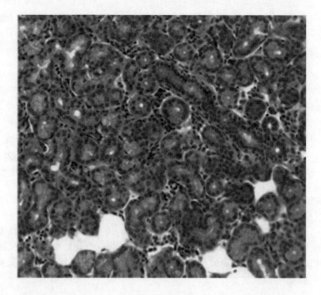

lacrimal gland cells with their apices pointed to a central lumen. The central lumen comprises many of these units and converges to form intralobular ducts. Intralobular ducts subsequently unite to form interlobular ducts. The lacrimal gland is similar in structure to the parotid gland with the lack of striated ducts.

9.2.2 Accessory Lacrimal Glands

9.2.2.1 Wolfring Glands
Wolfring glands (Ciaccio's glands) are small tubular accessory lacrimal glands found in the *lacrimal caruncle* of the eyelid. They are located in the upper border of the tarsus, approximately in the middle between the extremities of the tarsal glands. They are situated slightly above the tarsus. There are usually 2–5 of these glands in the upper eyelid and usually two glands in the lower lid. Their function is to produce tears, which are secreted onto the surface of the conjunctiva. Histologically, the glands are similar to the main lacrimal gland. Wolfring's glands are serous glands that contribute much of the proteinaceous components of the tear film including antibacterial agents such as lysozyme.

9.2.2.2 Krause's Glands
Krause's glands are an additional type of accessory lacrimal gland which are smaller and more numerous than Wolfring's glands. They are found along the superior and inferior *fornix* of the *conjunctival sac* and their ducts unite into a long sinus tract, which open into the *fornix conjunctiva*. There are approximately 40 Krause's glands in the region of the upper eyelid, and around 6–8 in the region of the lower lid. The

main function of Krause's glands is to produce tears, which are secreted onto the surface of the conjunctiva. Their histology is similar to the lacrimal gland.

9.2.3   Additional Exocrine Glands

9.2.3.1 Glands of Zeis

Glands of Zeis (Fig. 9.6) are sebaceous glands located on the margin of the eyelid. These glands produce an oily secretion that flows through the excretory ducts of the sebaceous lobule into the middle portion of the hair follicle.

9.2.3.2 Goblet Cells

Goblet cells are at the margin of the lid and the mucocutaneous junction of the outer epidermis and the conjunctiva of the eyelid. Goblet cells are located among the stratified columnar epithelium layer and contribute mucous secretions to the tear film (Fig. 9.7).

Fig. 9.6 The histological representation of the glands of Zeis. These glands produce an oily secretion which is indicated by the letter A. The excretory ducts are indicated by the letter B within the slide

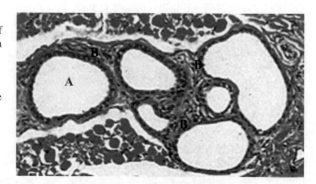

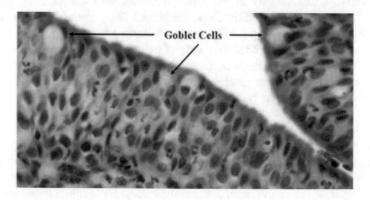

Fig. 9.7 The histological slide of the presence of goblet cells in the eyelid

9.2.3.3 Glands of Moll

Glands of Moll are modified apocrine sweat glands that are found next to the base of the eyelashes, and anterior to the Meibomian glands within the distal eyelid margin. They are relatively large and tubular-shaped and empty into the adjacent lashes.

9.2.3.4 Meibomian Glands

Meibomian glands (Fig. 9.8) are holocrine exocrine glands located along the borders of the eyelid within the tarsal plate. They produce *meibum*, a substance that prevents evaporation of the tear film. The glands are numerous, with approximately 50 glands on the upper eyelid and 25 glands on the lower eyelid. The most significant of these glands are the tarsal glands. They are linear arrays of sebaceous glands connected to central ducts that open at the mucocutaneous junction of the eyelid. There are approximately 20 glands in each of the upper and lower eyelids, and their sebaceous secretion contributes a critical lipid component to the tear film. The lipid component reduces surface tension and retards evaporation.

9.2.3.5 Crypts of Henle

Crypts of Henle are microscopic invaginations that are found in scattered sections of the conjunctiva round the eyeball. They are often referred to as pseudo glands because the invaginations contain numerous goblet cells among the pseudostratified columnar lining of the epithelia (Fig. 9.9). Goblet cells are responsible for the mucinous secretions that make up the inner layer of tears. It also coats the cornea to provide a hydrophilic layer that allows for even distribution of the tear film, which in turn allows the tears to glide evenly across the eye's surface.

9.2.3.6 Glands of Manz

Glands of Manz are located within the eyeball's bulbar conjunctiva near the limbus. They are arranged in a ring around the cornea, near the scleral junction. Like the crypts of Henle, they also are responsible for secreting mucin into tears.

Fig. 9.8 The histological slide of the Meibomian glands (in cross-section). A typical Meibomian gland is circled on the slide

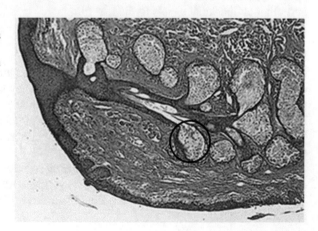

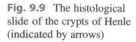

Fig. 9.9 The histological
slide of the crypts of Henle
(indicated by arrows)

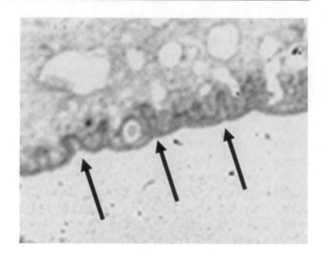

9.2.4   Ocular Exocrine Secretions

9.2.4.1 Tears

The palpebra functions primarily to protect the eye from mechanical damage and to help adjust the amount of incoming light. In addition, blinking the eyelid helps spread a liquid film of tears over the cornea (a structure discussed below), which, being avascular, is critically dependent upon the tear film as a means of gas exchange. Tears also contain lysozyme and other antibacterial proteins that help protect the cornea and conjunctiva from infection.

9.2.4.2 Sebum

Sebum can be characterized as a light yellow, oily substance that helps keep the skin and hair moisturized. *Sebum* contains triglycerides, free fatty acids, wax esters, squalene, cholesterol esters, and cholesterol. It may also contribute sIgA, lysozyme, and other antibacterial agents to the tear film.

9.2.4.3 Mucin

Mucins are high-molecular-weight glycoproteins with a protein backbone and a high carbohydrate content found on the ocular surface. They are primarily produced by goblet cells and the lacrimal gland. One role of mucins is to make the tear film hydrophilic, a factor that stabilizes the tear film and decreases its surface tension. Without this layer, tears would not adhere to the surface, making the ocular surface susceptible to damage.

   Goblet cells primarily produce the gel-forming secretory mucin MUC5AC, while the apical cells of the conjunctiva and cornea produce MUC1, MUC2, and MUC4. These mucins primarily protect the ocular surface. Finally, the lacrimal gland produces MUC7.

9.2.4.4 Meibum

Meibum, which is significantly different from sebum is a lipid-rich, oily substance that is constantly released from the orifices of the Meibomian glands. Lipids are universally recognized as major components of meibum; however, more than 90 different proteins have been identified in the Meibomian gland secretions. Overall, meibum prevents evaporation of the eye's tear film and traps the tears between the oiled edge and the eyeball thereby making the closed lids airtight.

9.2.5 Summary

Table 9.1 summarizes the exocrine glands of the eye and their contributions to the ocular tear film.

9.3   Glands of the Nose and the Tongue

The glands of the nose and the tongue have been previously discussed in Chaps. 3 and 4, respectively.

| Questions |

1. The auditory meatus has no eccrine sweat glands. Is this statement true or false?
   a. True
   b. False
2. List the three types of exocrine glands of the ear.
3. Explain the function of sebum.
4. List the types of exocrine glands of the eye.
5. Which one of the following is the largest gland of the eye:
   a. Moll
   b. Lacrimal
   c. Manz
   d. Wolfring
6. Explain the function and composition of tears.
7. The lacrimal glands, like the salivary glands, have striated ducts. Is this statement true or false?
   a. True
   b. False
8. Which one of the following of the exocrine glands or cells of the eye is serous:
   a. Moll
   b. Goblet
   c. Manz

      d. Wolfring
      e. Glands of Zeis
  9. Glands of Manz are located within the eyeball's bulbar conjunctiva near the limbus. Is this statement true or false?
      a. True
      b. False
 10. The glands of Moll are modified apocrine sweat glands. Is this statement true or false?
      a. True
      b. False

Suggested Reading

Junqueira, Luis, and Carneiro, Jose. 1983. "Sensory Organs". In *Basic Histology*, 4th ed., edited by Luis Junqueira and Jose Carneiro, 196–225. Los Altos: Lange Medical Publications.

Kelly, Douglas, Wood, Richard, and Enders, Allen. 1984. "The Organs of Special Senses". In *Bailey's Textbook of Microscopic Anatomy*, 18th ed., edited by Douglas Bailey, Richard Woods, and Allen Enders, 822–869. Baltimore: Williams & Wilkins.

Lobitz, Walter, and Campbell, Clarence. 1952. "Physiology of the Human Ear Canal: Preliminary Report". *Journal of Investigative Dermatology*, Vol. 19, No. 2: 125–135. https://doi.org/https://doi.org/10.1038/jid.1952.77.

Pawlina, Wojciech. 2016. "Eye". In *Histology: A Text and Atlas with Correlated Cell and Molecular Biology*, 7th ed., edited by Wojciech Pawlina, 900–903. Philadelphia: Wolters Kluwer Health.

Perry, Eldon, and Shelley, Walter. 1955. "The Histology of the Human Ear Canal with Special Reference to the Ceruminous Gland". *Journal of Investigative Dermatology*, Vol. 25, No. 6: 439–451. https://doi.org/https://doi.org/10.1038/jid.1955.149

Van Haeringen, Nichlaas. 1997. "Aging and the Lacrimal System". *British Journal of Ophthalmology*, Vol. 81: 824–826.

Answers to Chapter Questions

Chapter 1

1. Define glands.
   Answer: Glands by definition are composed of distinct types of cells, which are specialized to produce substances to be used elsewhere in the body. These cells are known as the glandular epithelium. These cells form an aggregate and develop into an organized structure for secretion or excretion. The aforementioned aggregations are known as glands. Glands are classified according to their secretory system. In the human body, there are two types of glands: *endocrine* and *exocrine*. The study of endocrine glands is called *endocrinology* and the study of the exocrine glands is called *exocrinology*.

2. Describe the difference between endocrine and exocrine glands.
   Answer: Endocrine glands are composed of specialized epithelial cells that secrete their products directly into the blood or lymph. The exocrine glands, however, open to the surface (e.g., the sweat glands) or into a lumen, which eventually connects to the body surfaces such as the gastrointestinal tract or the lungs.

3. Describe the different types of exocrine glands.
   Answer: The major glands of the exocrine system are the lacrimal glands, mammary glands, sweat glands, salivary glands, the exocrine pancreas, the kidney, and the liver; however, there is a multitude of minor exocrine glands throughout the body.

4. Name the two types of simple exocrine glands.
   Answer: There are two types of simple exocrine glands: the *tubular* and the *alveolar glands*.

5. What are the three secretory methods among exocrine glands?
   Answer: There are three secretory methods among exocrine glands: *merocrine*, *apocrine*, and *holocrine*.

© The Author(s), under exclusive license to Springer Nature Switzerland AG 2022   119
C. F. Streckfus, *Exocrinology*, https://doi.org/10.1007/978-3-030-97552-4

Chapter 2

1. List the layers of the integument?
   Answer: epidermis and dermis
2. Name the four types of exocrine glands of the integument.
   Answer: stratum basale, stratum spinosum, stratum granulosum, stratum lucidum, stratum corneum
3. Which statement(s) is true about the sebaceous glands?
   Answer: e. All of the above
4. The integument has both eccrine and apocrine glands. Is the statement true or false?
   Answer: True
5. Which of the following is not a layer of the epidermis.
   Answer: c. Stratum various
6. The apocrine glands are activated during puberty. Is the statement true or false?
   Answer: a. True
7. Apoeccrine glands secrete more perspiration than both eccrine and apocrine glands; thus, they play a large role in axillary sweating. Is the statement true or false?
   Answer: a. True
8. Which of the four exocrine glands is not coiled tubular in shape?
   Answer: d. sebaceous glands.

Chapter 3

Questions

1. Which portion of the respiratory tract are the Bowman's glands located?
   Answer: Nasal cavity
2. Club cells were originally called *Clara cells*, but the name has been changed due to the unethical methods used by Max Clara in obtaining human respiratory tissue. Is this statement true or false?
   Answer: a. True
3. Goblet cells occur in all of the following except:
   Answer: d. Alveoli
4. Which of the following is MISMATCHED?
   Answer: d. pulmonary surfactant--increases surface tension within alveoli.
5. Which type of cell produces surfactant? Type I or Type II pneumocytes?
   Answer: Type II pneumocytes
6. Gaseous exchange in the alveoli occurs across the endothelial cell of the capillary, and which type of pneumocyte?
   Answer: Type I pneumocytes

Chapter 4

1. The Von Ebner's glands are exocrine glands found on the tongue. More specifically, they are serous salivary glands which assist in the tasting of food. Are these statements true or false?
   Answer: a. True
2. All the following statements regarding saliva buffering capacity are True EXCEPT:
   Answer: b. A low pH is vital to the health of the dentition
3. Which of the following are true about salivary amylase?
   Answer: b. Amylase initiates digestion in the oral cavity.
4. Which statement(s) are True about lingual lipase?
   Answer: e. All the above
5. Which salivary analyte(s) ARE NOT involved in the mineral composition & maintenance of enamel?
   Answer: d. Secretory IgA
6. Which statement(s) are true about Statherin?
   Answer: c. Amino-terminal hexapeptide inhibits secondary precipitation (crystal growth)
7. Acinar epithelium secretes proteins and hypotonic filtrate. Is this statement true or false?
   Answer: b. False
8. Which statement is false concerning the about the submandibular gland?
   Answer: a. The submandibular salivary gland, like the parotid gland, is serous.
9. Which statement is false concerning salivary secretions?
   Answer: e. None of the above
10. Parasympathetic controls increases salivary fluid secretion while sympathetic innervation of the salivary gland increases protein secretion. Is this statement true or false?
    Answer: a. True
11. A patient is given an anticholinergic drug for the treatment of arrhythmias (irregular heart beat). Would you expect the medication to:
    Answer: b. Decrease salivary flow
12. A patient is given an antiadrenergic drug for the treatment of hypertension. Would you expect the medication to:
    Answer d. Decrease protein secretion
13. Which of the following are true about the secretion of saliva in the acinar region?
    Answer: e. All the above
14. Passive diffusion, active transport and endocytosis/exocytosis are all mechanism by which particulates can enter into saliva. Is this statement true or false?
    Answer: a. True
15. Which of the following salivary proteins does not have a role in host defenses?
    Answer: d. Amylase
16. Serum and salivary IgA are exactly the same. Is this statement true or false?
    Answer: b. False

17. Both IgA1 and IgA2 have been found in external secretions like maternal milk, tears and saliva. Is this statement true or false?
    Answer: a. True

Chapter 5

1. The esophageal glands are tubular alveolar glands that secrete mucous at the superior portion of the esophagus. Is this statement true or false?
   a. True
   b. False
2. Sebaceous glands are primarily located in which part of the gastrointestinal tract.
   Answer: e. anus
3. The esophageal glands are tubular alveolar glands that secrete mucous at the superior portion of the esophagus. Is this statement true or false?
   a. True
   b. False
4. Brunner's glands are primarily located in which part of the gastrointestinal tract:
   Answer: c. small intestines
5. The Chief cells secrete pepsin, Is this statement true or false?
   Answer: b. False
6. Which exocrine glands listed below are compound tubular glands?
   Answer: f. Brunner's and pyloric
7. The parietal cells secrete HCl and are located in the large intestines. Is this statement true or false?
   Answer: b. False
8. Some anal glands produce pheromones. Is this statement true or false?
   Answer: a. True
9. Cobelli's glands are primarily located in which part of the gastrointestinal tract:
   Answer: a. esophagus
10. The Lieberkühn glands secrete mucous, tubular in structure and are found in the esophagus. Is this statement true or false?
    Answer: b. False

Chapter 6

1. The pancreas has striated ducts. Is this statement true or false?
   Answer: b. False
2. Which of the following is a compound tubuloacinar serous organ?
   Answer: a. Pancreas
3. Which of the following have a cellular structure similar to the salivary glands?
   Answer: a. Pancreas
4. Both the salivary glands and the pancreas secrete amylase. Is this statement true or false?

Answer: a. True
5. Which of the following organs contain peribiliary tubuloalveolar mucous glands?
   Answer: c. Gall bladder
6. The space of Disse is found in which of the following organs?
   Answer: b. Liver

Chapter 7

1. The Skene's cells produce a serous secretion. Is this statement true or false?
   Answer: b. False
2. Which of the following is a compound tubuloacinar gland?
   Answer: a. Mammary gland
3. The organs of reproduction can be considered both exocrine and endocrine. Is this statement true or false?
   Answer: a. True
4. All of the following glands are mucinous except one? Identify the non-mucinous gland.
   Answer: a. Mammary gland
5. The internal glands of the female reproductive system are primarily mucinous. Is this statement true or false?
   Answer: a. True
6. Which one of the following compounds is produced primarily by the seminal vesicles?
   Answer: b. Ergothioneine
7. The non-ciliated secretory cells, also known as peg cells, release a secretion that lubricates the oviduct and provides nourishment and protection to the traveling ovum. Is this statement true or false?
   Answer: a. True
8. The Bartholin's glands are compound alveolar glands located slightly posterior and to the left and right of the opening of the vagina. Is this statement true or false?
   Answer: a. True

Chapter 8

1. There are numerous exocrine glands associated with the ureters and the bladder. Is this statement true or false?
   Answer: b. False
2. Urine contains which of the following compounds?
   Answer: e. a, b, c
3. The kidney is a compound tubular gland. Is this statement true or false?
   Answer: a. True
4. Describe the difference between secretion and excretion.

Answer: Secretion produces products to be used within the body where excretion for the removal of waste products from the body.

5. The distal convoluted tubule has smaller epithelia and does not have microvilli. Is this statement true or false?

Answer: a. True

Chapter 9

1. While the auditory meatus has no eccrine sweat glands. Is this statement true or false?
   Answer: a. True

2. List the three types of exocrine glands of the ear.
   Answer: There are three types of glands in the external ear: sebaceous, eccrine sweat (merocrine sudoriferous), and ceruminous (wax) glands.

3. Explain the function of sebum.
   Answer: *Sebum* is a light yellow, oily substance that helps keep the skin and hair moisturized.

4. List the types of exocrine glands of the eye.
   Answer: There are pair of lacrimal glands and seven additional exocrine glands: glands of *Henle, Krause, Meibomian, Manz, Moll, Wolfring (Ciaccio),* and *Zeis* (Table 9.1). There are also abundant goblet cells.

5. Which one of the following is the largest gland of the eye.
   Answer: b. Lacrimal

6. Explain the function and composition of tears.
   Answer: Tears cover the cornea which, being avascular, is critically dependent upon the tear film as a means of gas exchange. Tears also contain lysozyme and other antibacterial proteins that help protect the cornea and conjunctiva from infection.

7. The lacrimal glands, like the salivary glands, have striated ducts. Is this statement true or false?
   Answer: b. False

8. Which one of the following of the exocrine glands or cells of the eye is serous.
   Answer: d. Wolfring

9. Glands of Manz are located within the eyeball's bulbar conjunctiva near the limbus. Is this statement true or false?
   Answer: a. True

10. The glands of Moll are modified apocrine sweat glands. Is this statement true or false?
    Answer: a. True

Index

Printed in the United States
by Baker & Taylor Publisher Services